M000308815

B.C. Science PROBE  6

AUTHORS

Susan Doyle
Curriculum Developer and Author
Writing Instructor, University of Victoria

Jean Bowman
Saanich School District, #63

Darlene Vissers
Abbotsford School District, #34

PROGRAM CONSULTANT

Arnold Toutant
Educational Consultant, A. Toutant Consulting Group Ltd.

THOMSON

NELSON

Australia   Canada   Mexico   Singapore   Spain   United Kingdom   United States

THOMSON

NELSON

B.C. Science Probe 6

Authors
Susan Doyle
Jean Bowman
Darlene Vissers

Program Consultant
Arnold Toutant

Associate Vice President of Publishing:
David Steele

**Senior Publisher,
Assessment and Science:**
Bill Allan

Acquisitions Editor, Science:
John Yip-Chuck

**Executive Managing Editor,
Development:**
Cheryl Turner

Project Manager, Science:
Lois Beauchamp

Program Manager:
Lee Geller

Project Editor:
Lee Ensor

Developmental Editors:
Julie Bedford, Janis Barr

Editorial Assistant:
Alisa Yampolsky

Executive Managing Editor, Production:
Nicola Balfour

Senior Production Editor:
Linh Vu

Copy Editor:
Paula Pettitt-Townsend

Proofreader:
Gilda Mekler

Indexer:
Noeline Bridge

Senior Production Coordinator:
Sharon Latta Paterson

Creative Director:
Angela Cluer

Art Director:
Ken Phipps

Text Design:
Kyle Gell Design, Peter Papayanakis,
Ken Phipps

Art Management:
Allan Moon, Suzanne Peden

Composition Team:
Kyle Gell, Allan Moon

Cover Design:
Peter Papayanakis

Cover Image:
Michael Patrick O'Neill
Photo Researchers Inc.

Illustrators:
Steven Corrigan
Deborah Crowle
Margo Davies LeClair
Kyle Gell
Irma Ikonen
Imagineering
Kathy Karakasidis
Dave Mazierski
Dave McKay
Allan Moon
Bart Vallecoccia
Cynthia Watade
Dave Whamond

Photo Research and Permissions:
Mary Rose MacLachlan

Printer:
Transcontinental Printing Inc.

Library and Archives Canada
Cataloguing in Publication Data

Doyle, Susan, 1952–
B.C. science probe 6 / Susan Doyle, Jean
Bowman, Darlene Vissers.

Includes index.
ISBN 0-17-627176-7

1. Science—Textbooks. I. Bowman, Jean
(Jean Elizabeth) II. Vissers, Darlene III.
Title. IV. Title: B.C. science probe six.

Q161.2.B69 2005   500   C2004-
905928-9

REVIEWERS

Aboriginal Education Consultant
Mary-Anne Smirle
Peachland Elementary School, Central Okanagan
    School District (#23), B.C.

Accuracy Reviewers
Gordon Gore
Formerly of Kamloops/Thompson School
    District (#73), B.C.
    Operator, BIG Little Science Centre, Kamloops

Brian Herrin
Simon Fraser University, B.C.

Dr. Fiona McLaughlin
Research Scientist, Fisheries and Oceans Canada

Assessment Consultant
Anita Chapman
Educational Consultant and Author, Chapman &
    Associates Educational Consulting, Inc.

ESL Consultant
Vicki McCarthy, Ph. D.
Dr. George M. Weir Elementary School, Vancouver
    School District (#39), B.C.

Literacy Consultant
Sharon Jeroski
Educational Consultant and Author,
    Horizon Research and Evaluation, Inc.

Numeracy Consultant
Bob Belcher
Happy Valley at Metchosin Elementary School,
    Sooke School District (#62), B.C.

Professional Development Consultant
Brian Herrin
Simon Fraser University, B.C.

Safety Consultant
Marianne Larsen
Formerly of Sunshine Coast School
    District (#46), B.C.,
    Greater Vancouver Regional Science Fair

Technology Consultant
Al Mouner
Millstream Elementary School, Sooke School
    District (#62), B.C.

REVIEWERS

Advisory Panel and Teacher Reviewers

Doug Adler
Department of Curriculum Studies, Science Education, University of British Columbia

Cheryl Angst
Minnekhada Middle School, Coquitlam School District (#43), B.C.

Wade Blake
Rutherford Elementary School, Nanaimo-Ladysmith School District (#68), B.C.

Margaret Boyd
Anne McClymont Elementary School, Central Okanagan School District (#23), B.C.

George Clulow
Literacy/Intermediate Science Consultant, Langley School District (#35), B.C.

Burt Deeter
James Ardiel Elementary School, Surrey School District (#36), B.C.

Pat Horstead
Mount Crescent Elementary School, Maple Ridge-Pitt Meadows School District (#42), B.C.

Jillian Lewis
Lyndhurst Elementary School, Burnaby School District (#41), B.C.

Susan Martin
Cliff Drive Elementary School, Delta School District (#37), B.C.

Ann McDonnell
Blewett School, Kootenay Lake School District (#8), B.C.

Darlene Monkman
Formerly of Nanaimo-Ladysmith School District (#68), B.C.

Karen Morley
North Surrey Learning Centre, Surrey School District (#36), B.C.

Noreen Morris
Trafalgar Elementary School, Vancouver School District (#39), B.C.

Len Reimer
Sir Wilfrid Laurier Elementary School, Vancouver School District (#39), B.C.

Carollyne Sinclair
Faculty Associate, Simon Fraser University, B.C.

Mary-Anne Smirle
Peachland Elementary School, Central Okanagan School District (#23), B.C.

Heather Stannard
Khowhemun Elementary School, Cowichan Valley School District (#79), B.C.

Patricia Tracey
Formerly of Abbotsford School District (#34), B.C., Fraser Valley Regional Science Fair Foundation

Kyme Wegrich
Glen Elementary School, Coquitlam School District (#43), B.C.

CONTENTS

UNIT B: ELECTRICITY

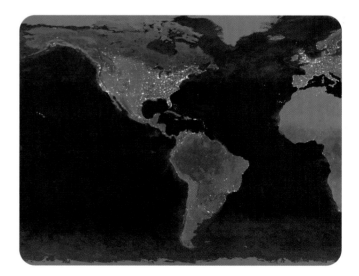

UNIT C: EXPLORING EXTREME ENVIRONMENTS

EXPANDING THE WORLD OF SCIENCE

What is a scientist?

Grade 6 students at Peachland Elementary School were asked this question. This is how some students responded:

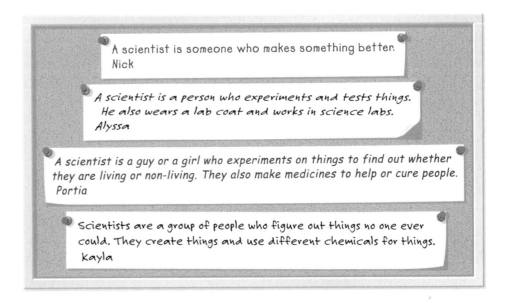

> A scientist is someone who makes something better.
> Nick

> A scientist is a person who experiments and tests things. He also wears a lab coat and works in science labs.
> Alyssa

> A scientist is a guy or a girl who experiments on things to find out whether they are living or non-living. They also make medicines to help or cure people.
> Portia

> Scientists are a group of people who figure out things no one ever could. They create things and use different chemicals for things.
> Kayla

The dictionary defines a scientist as someone who is an expert in a science. It is true that many scientists do experiments in a lab, but many others develop theories and study the world in its natural environment as well. One thing that all scientists have in common is that they share what they have learned with others. This is a very important part of science.

In Canada, First Nations and Inuit peoples have lived in their traditional territories long before the first explorers and immigrants arrived in North America. These Indigenous groups, and later the Métis people, lived very closely with nature. They developed very detailed knowledge about the places they lived in and about their environment. This includes knowledge about plants, animals, weather, and landforms. Like all scientists, this information was shared from one generation to the next, within the Aboriginal community, so that the knowledge wouldn't be lost.

Now this information, sometimes known as **Indigenous Knowledge** (IK) or Traditional Ecological Knowledge (TEK), is being shared outside the Aboriginal community. In British Columbia, like in other places all around the world, more and more scientists are asking to work with and learn from Aboriginal peoples and scientists. When this happens it can be a powerful partnership for everyone involved. It is important to Aboriginal peoples that their ways are respected and valued by others. We can all learn from the valuable information that Aboriginal peoples have to share.

Mary-Anne Smirle
Métis Nation
L'Hirondelle clan

UNIT A

THE DIVERSITY OF LIFE

CHAPTER **1** Living things have similarities and differences.

CHAPTER **2** Classifying living things helps us understand the diversity of life.

CHAPTER **3** Magnifying tools make the invisible world visible.

CHAPTER **4** Living things adapt to their environments.

Preview

Welcome to life—the most amazing show on Earth! Our planet is crawling, swimming, hopping, and buzzing with living things. They come in every colour, shape, and size you can imagine, from giant trees that tower over us to tiny bacteria that can only be seen with a microscope. There is life on barren mountaintops, in sunbaked deserts, in the deepest oceans, and in the icy waters of the Antarctic Ocean. In fact, there isn't any place on Earth where life doesn't exist. Even more amazing, Earth is the only place where life does exist—as far as we know.

Look at the photo on this page. How many different living things can you see? Why does life come in so many different forms? Do you know what makes all forms of life alike in some ways? Do you know what makes them like you?

In this unit, you will discover the answers to these questions and many other questions about living things. You will learn how scientists observe living things and classify them into groups. Like a scientist, you will use a microscope to investigate living things that are too tiny for just your eyes to see. As you do these activities, you will follow the same steps that scientists follow as they explore the incredible diversity of living things. What you discover might surprise you!

TRY THIS: MAKE A LIVING WORLD WEB

Skills Focus: questioning, predicting

1. Create a web that shows what you already know about the variety of life on Earth. Write the words "Life on Earth" in the centre. Radiating from the centre, write different questions about the diversity of life on Earth. For example, you could include questions such as these: What does it mean to be living? What kinds of living things exist? How do living things adapt to survive? What do I know about living things?

2. Beside each question, write a "best guess" answer. Do not worry about being correct. Use your imagination!

◀ A tidal pool in Burnaby Narrows, Queen Charlotte Islands, British Columbia, contains a variety of life like these colourful sea stars.

Living things have similarities and differences.

KEY IDEAS

- ▶ Living things share characteristics.
- ▶ Living things share basic needs.
- ▶ Living things are diverse.

It is easy to see how the dolphins and the snorkelling child in the photo are different from each other. The dolphins are in their natural habitat, while the child is using fins and a snorkel to swim in the water. But how are they the same? The dolphins and the child are just two examples of the incredible diversity of living things on Earth. In this chapter, you will discover that dolphins and children, as well as every other living thing, are surprisingly alike in many ways.

The Characteristics of Living Things

The world around us is made up of both living and non-living things. But how do we tell the difference? Look at **Figure 1**. Which things are living and which are non-living? Are the rocks living things? Is the lake itself a living thing? What is the difference between the twigs growing on the tree's branches and the twigs in the bird nest?

Figure 1
What living things do you see in this picture? What non-living things do you see?

LEARNING TIP ◁

You have already studied living things in earlier grades. Look at the headings in this chapter and review what you know about the similarities among living things.

One way to identify living things, or **organisms,** is to look at the characteristics they have in common.

Living things

- are made of one or more cells
- grow and develop
- reproduce
- respond

Non-living things, such as rocks and buildings, do not have these characteristics.

LEARNING TIP ◁

Important vocabulary words are highlighted. These are words that you should learn and use when you answer questions. These words are also defined in the glossary at the back of this book.

Living things are made of one or more cells. A cell is a tiny, microscopic structure that is the basic unit of all living things. Some living things, such as the bacteria shown in **Figure 2**, are made up of only one cell. Other living things, such as the deer shown in **Figure 3**, contain many cells. You are made of trillions of cells.

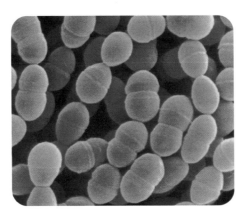

Figure 2
Bacteria

Figure 3
Deer

Living things grow and develop. Some organisms, such as a butterfly, change shape as they grow and develop. **Figure 4** shows the life cycle of a butterfly. Other organisms, such as a cat or a human, are born looking like miniature adults. Most organisms have a life span—the maximum time that they can live. Some bacteria live for only a few hours. A mayfly's life span is one to three days, and a human's life span is over 110 years. Some plants and fungi can live for more than 10 000 years!

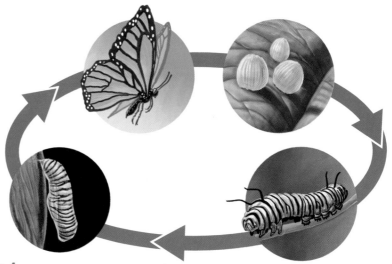

Figure 4
A caterpillar grows and develops into a butterfly.

Living things reproduce to form more of the same kind of organism. These offspring are also able to reproduce. There are many methods of reproduction. Some organisms, such as humans, give birth to live young. Other organisms, such as birds and fish, hatch from eggs. **Figure 5** shows a bald eagle nesting in a tree. Plants develop from seeds or spores. Bacteria reproduce by splitting into two identical cells.

Figure 5
A bald eagle

Living things respond. For example, if you touch something hot, you respond by quickly pulling your hand away. Or if you look into a bright light, you respond by squinting your eyes. Some plants, such as sunflowers, respond to light by turning towards it (**Figure 6**).

Figure 6
Sunflowers take their name from the way they turn to face the Sun.

▶ CHECK YOUR UNDERSTANDING

1. Look at **Figure 1**. What is the difference between the fish in the lake and the fish in the net?

2. Which main characteristic of living things does each statement represent?
   - You are constantly losing skin cells.
   - A rabbit gives birth to babies.
   - Lobsters can live for 50 years.
   - A tadpole develops into a frog.
   - Bacteria divide to form two identical, smaller cells.
   - A sow bug rolls itself into a ball when it is touched.

3. Choose an organism that lives in your community. Use the characteristics of living things to show that it is a living thing.

1.1 The Characteristics of Living Things

1.2 The Needs of Living Things

Figure 2
Aphids suck the sap from plants.

All organisms have the same basic needs. They must find these things within their environments, or they will die.

Living things need nutrients and energy. Nutrients are substances that organisms need to keep healthy and grow. Nutrients are found in foods and in the soil. Organisms also need energy to grow and develop, and to reproduce. Different organisms use different types of energy. Plants use the Sun's energy to make their own food. The aphids in **Figure 2** feed on plants. Spiders and birds eat the aphids. In this way, the Sun's energy is passed from one living thing to another.

Living things need water. Water is the main ingredient of the cells of all living things. You, for example, are about two-thirds water! Without water, you could live for only a few days. Some organisms, such as the cactus in **Figure 3**, can live in a very dry environment by storing moisture in their stems or leaves.

Figure 3
A prickly pear cactus in Fraser Canyon, British Columbia.

Living things need air. You, and other organisms that live on land, get oxygen from the air. Fish use oxygen in the water. Marine mammals, such as the orca in **Figure 4**, come to the surface for oxygen. Green plants use carbon dioxide, water, and sunlight to make food.

Figure 4
The orca breathes air at the surface.

Living things need a habitat, or a place to live. A habitat may be a hole in a tree or an isolated mountaintop (**Figure 5**). It is a place where an organism has living space and the right conditions, such as nutrients and energy, water, air, and temperature, to survive. Usually, many organisms share a habitat. For example, coral reefs provide a habitat for many plants, animals, and other marine organisms.

Figure 5
The habitat of the mountain goat includes steep cliffs and rocky slopes.

▶ CHECK YOUR UNDERSTANDING

1. In your notebook, make a table like the one below. Use what you have learned about the needs of organisms to complete your table.

Need of organism	Two examples of how organisms meet this need
nutrients and energy	- Plants get energy from the Sun. -
water	
habitat	

2. How do you meet each of the basic needs listed in the table? (For example, you get energy from the food you eat.)

1.2 The Needs of Living Things

1.3 Design Your Own Experiment

SKILLS MENU

- ○ Questioning
- ● Observing
- ○ Predicting
- ● Measuring
- ● Hypothesizing
- ○ Classifying
- ● Designing Experiments
- ● Inferring
- ● Controlling Variables
- ● Interpreting Data
- ○ Creating Models
- ● Communicating

▷ **LEARNING TIP**

For help with this activity, read the Skills Handbook sections "Designing Your Own Experiment," "Hypothesizing," and "Controlling Variables."

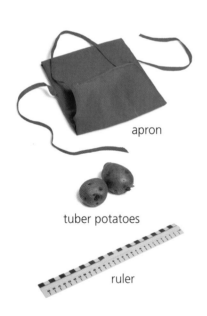

apron

tuber potatoes

ruler

What Factors Affect the Growth of a Potato?

All plants need energy from the Sun, water, and a habitat that provides the right temperature to live. Each of these needs is a factor, or variable, in the plant's survival. Design an experiment to test how sunlight, water, and temperature affect the growth of two potatoes.

Question

How does the amount of energy from the Sun, water, or temperature affect the growth of potatoes?

Hypothesis

Write a hypothesis that answers the question. Make sure that you complete your hypothesis with a short explanation of your reasons. Write your hypothesis in the form "If . . . then . . . because . . . ."

Materials

- apron
- 2 tuber potatoes
- ruler

Decide what other materials you will need. Check with your teacher to make sure that these materials are safe for you to use.

- Design a procedure to test your hypothesis. A procedure is a step-by-step description of how you will conduct your experiment. It must be clear enough for someone else to follow your instructions and do the exact same experiment.

- Submit your procedure, including any safety precautions, to your teacher for approval. Also submit a diagram, at least half a page in size, showing how you will set up your experiment.

Data and Observations

Create a table to record your observations. Record your observations as you carry out your experiment.

Analysis

1. Describe the growth of the potato that received the lesser amount of the variable you tested. How well did it grow?

2. Describe the growth of the potato that received the greater amount of the variable you tested. How well did it grow?

3. How does the variable you tested affect the growth of potatoes?

Conclusion

Look back at your hypothesis. Did your observations support, partly support, or not support your hypothesis? Write a conclusion that explains the results of your experiment.

Applications

1. How could you use what you learned from your experiment when growing plants at home?

2. Why would your conclusions be important information for a garden store or for a grocery store that doesn't want the potatoes to grow?

LEARNING TIP ◁

When you make observations, it is important to be accurate and complete. Read the Skills Handbook section "Observing" to learn about the different types of observations and why each one is important.

▌▶ *CHECK YOUR UNDERSTANDING* ⊗

1. How did your understanding of the needs of living things help you form a hypothesis for your experiment?

2. What was the independent variable in your experiment? What was the dependent variable?

3. Why was it important to change only one variable?

Living Things Are Diverse

You have learned that all living things are alike in some ways. They share certain characteristics and they have the same basic needs. But aside from these similarities, living things come in an astonishing variety of forms (**Figure 1**). In fact, the most amazing thing about life is the variety, or diversity, of living things on Earth.

Figure 1
What are some of the differences between the scarlet macaw and the whale shark?

For example, living things come in all sizes, from the towering giant sequoia (**Figure 2**) to organisms that are so small they cannot be seen with the naked eye (**Figure 3**). Magnifying tools, such as microscopes, have allowed scientists to identify thousands of tiny organisms, including some that live on and inside our bodies. You will use magnifying tools to look at living things in Chapter 3.

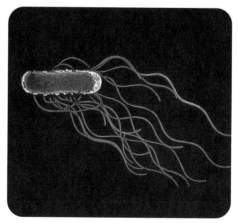

Figure 2
The largest giant sequoias are as tall as a 26-storey building.

Figure 3
A bacterial cell can only be seen under a microscope.

Organisms also get their food in different ways. Plants can make their own food. They use sunlight to turn carbon dioxide and water into food in a process called photosynthesis [foh-toh-SIN-thuh-sis]. However, there are some plants, like the Venus flytrap, that capture and eat small insects (**Figure 4**). Fungi, such as mushrooms, live right on their food source. Animals have to look for their food sources. For example, think about how you get your food. You have to search for your food, even if it's just in the kitchen. In general, animals eat plants, other animals, or the remains of living things (**Figure 5**).

Figure 4
The Venus flytrap is able to make its own food and feed on insects that it captures.

Figure 5
The great blue heron eats fish, turtles, and frogs.

LEARNING TIP ◁

Look at the three pairs of photos in this section. Think about what is being compared in each pair of photos.

Organisms move in every imaginable way. Some move along on legs (**Figure 6**). Those with wings, such as robins and bats, are able to fly overhead. Some organisms, such as fish and marine mammals, swim through the water using fins or flippers. Plants, on the other hand, remain in one place for their entire lives.

There is a great diversity of organisms on Earth and new organisms are discovered every day. Scientists have now identified over 1.7 million different organisms; however, they believe there may be more than 10 million organisms that have not yet been identified. In the next chapter, you will look at how all these diverse organisms can be organized.

gross

Figure 6
Although this giant desert centipede doesn't have 100 legs (as its name implies), it does have two legs on each of its body segments.

⫸ CHECK YOUR UNDERSTANDING ⊗

1. What are some of the ways that organisms differ from one another?

The living world is full of strange and wonderful creatures—and lots of surprises!

SPIDERS BY THE MILLION!

Scientists were amazed to discover a 24-hectare spider web covering a field of clover east of McBride, British Columbia. When they took a closer look, they found tens of millions of spiders (about two spiders per square centimetre), frantically engaged in the mystery building project. The scientists don't believe that the monster web was meant to be a giant insect trap because the spiders did not seem interested in the insects caught in it. But they still don't know what it was. One scientist joked that maybe the spiders were trying to catch a sheep! How could the scientists find out more about this strange phenomenon? What would you want to find out?

GREAT IMPOSTERS

With eyes like these, it's no wonder that the swallowtail caterpillar scares off predators. Or are they eyes? In fact, they're eyespots— markings that look like the eyes of a much larger creature. Eyespots are an example of mimicry. They are an effective way to keep from becoming someone's lunch!

The hawk moth is another great mimic. The snake-like appearance of the hawk moth caterpillar scares predators looking for a tasty feast. As a mature moth, its brown wings and shape easily blend into the bark of a tree. This makes it nearly impossible to see. Can you find other examples of organisms that use mimicry to survive?

REALLY SMALL-SCALE FARMING AND RANCHING

Humans are not the world's only farmers, or even the world's first farmers. In fact, the world's first farmers were leaf-cutter ants. These tiny ants cut out small pieces of leaves, chew them up, and leave them to decompose. The fungus that grows on the decaying mass is harvested and used as the ants' main food supply.

Other ants are ranchers, herding aphids onto young, sap-rich plants. The ants protect the aphids from predators to get their "milk"—the honeydew that the aphids excrete. The ants feed this milk to their young. What other characteristics do ants and humans share?

JUST GROW IT AGAIN!

What would you get if you cut a flatworm into four pieces? Wait two weeks and you'll have four new worms! A flatworm can regenerate, or grow back, a lost part of its body.

How does this happen? The secret is in the cells. A flatworm has stem cells that can be sent into action when its body has been damaged. Stem cells are not specialized. They can become any type of cell. For the flatworm, they develop into tissue for the other half of the worm's body. How could understanding regeneration be helpful to humans?

OPEN WIDE!

The leatherback turtle is the largest and most ancient species of sea turtle living today. Although it does not have any teeth, it can sure bite hard! It uses two upper "fangs"

to capture a jellyfish. Then it uses long, backward-facing spines in its mouth to swallow the meal. In fact, the leatherback's mouth protects it so well that it can eat a poisonous Portuguese man-of-war jellyfish without even getting stung!

GIANT DRAGONS DO EXIST!

It is 3 m long and has razor-like teeth and poisonous saliva. If that's not bad enough, it can run as fast as a dog for short distances. Fortunately, the Komodo dragon is found on only a few small islands in Indonesia. While it gets the name "dragon" from its fearsome characteristics, it is actually the world's largest lizard. In addition to its speed and its ability to spot objects up to 300 m away, it's the dragon's sense of smell that makes it so deadly. Whipping its long tongue in and out, the Komodo dragon samples the air and can find a meal 4 km away. What can you infer about the sense organs on the Komodo dragon's tongue?

Chapter Review

Living things have similarities and differences.

Key Idea: Living things share characteristics.

Vocabulary
organisms p. 5
cell p. 6

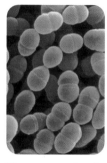

They are made of one or more cells.

They grow and develop.

They reproduce.

They respond to light.

Key Idea: Living things share basic needs.

Nutrients and energy

Water

Air

Habitat

Key Idea: Living things are diverse.

Review Key Ideas and Vocabulary

When answering the questions, remember to use the chapter vocabulary.

1. Create two columns on a piece of paper. Compare a salmon and a tomato plant, based on the characteristics of living things.

2. How is each of the following organisms meeting its needs?
   - Worms burrow into the ground.
   - Mudpuppies are solitary. They build walls around their territory to keep others away.
   - The Gila monster stores fat in its tail.
   - A Venus flytrap snaps shut when it senses an insect on its leaves.
   - A frog soaks up water through its skin.

3. What do scientists mean when they talk about the "diversity of life"?

Use What You've Learned

4. Volcanoes grow over time. Use what you have learned about the characteristics and needs of living things to explain whether or not volcanoes are living things.

5. Look at the photo below. Name an organism that could survive in this environment. How would the organism meet each of its needs?

6. When would it be important to understand the differences between living and non-living things?

Think Critically

7. List at least five benefits of living on a planet that has such a diversity of organisms.

Reflect on Your Learning

8. List three questions that you still have about living things. Glance through the rest of this unit. Do you think your questions will be answered in the topics that are covered? If not, where can you go to find the answers?

CHAPTER 2

Classifying living things helps us understand the diversity of life.

. 28

▥▶ KEY IDEAS

▶ People use classification systems to organize the diversity of living things.

▶ Living things can be unicellular or multicellular.

▶ Scientists classify organisms into groups based on internal and external features.

▶ Scientists classify living things into five kingdoms: Animalia, Plantae, Fungi, Protista, and Monera.

Have you ever hunted through your bedroom for a favourite T-shirt or CD? If you have, then you know how frustrating it is not to be able to find something you want. You also know that if you sorted things into groups, it would be easier to keep track of them. How are the things grouped in the above photo? How does each system help us find what we are looking for?

Scientists also sort things into groups so that they are easier to understand. There is such an incredible diversity of life on Earth. Living organisms come in all sizes, shapes, colours, and textures. How do scientists classify all these different organisms into groups?

Ways of Classifying Living Things

When you classify things, such as books or DVDs, you put the things that have similar characteristics together. These characteristics could be topic, author, or size. Sorting things into groups makes it easier to keep track of them.

Organizing things according to their similarities and differences is called classification. You used a **classification system** to sort living and non-living things in Chapter 1. You found that living things share the same characteristics and needs.

But how could you further classify all the thousands of living things around you? People use different systems, depending on what they want to know and what is important to them. For example, if you were on a deserted island, the first thing you would probably want to know is which plants you could eat and which plants you couldn't. Eventually, you would need and want to know much more. Grouping living things according to your observations helps you keep track of your knowledge.

TRY THIS: GROUP ORGANISMS

Skills Focus: observing, classifying

Look at the organisms in **Figure 1**. You can group them in many different ways. For example, you can group them by how they move, what they eat, where they live, and what they look like.

1. Explain how you would group these organisms.

2. What characteristics did you use to group these organisms?

Figure 1

Traditional Ways of Classifying Living Things

All people use classification systems to organize their knowledge of the living things around them. In the past, people relied on their detailed knowledge of living things to help them survive on the plants and animals that were available to them.

Aboriginal peoples, for example, use classification systems that are based on careful observation of the living world. The Aboriginal peoples of the northwest coast of British Columbia, for example, have classified over 200 different plants according to their uses, such as food and medicine (**Figure 2**). This information has been passed from generation to generation.

Figure 2
Some Aboriginal peoples use the leaves and twigs of the wild lilac plant to treat pain.

Aboriginal peoples also classify animals according to important characteristics. For example, they classify animals according to which animals are useful and which are dangerous, or where the animals are found. They also classify animals on the basis of helpful information, such as the season in which the animals can be hunted or the animals' use as a source of clothing or food.

Scientific Ways of Classifying Living Things

Scientists use classification systems to help understand the diversity of life on Earth. They examine the internal structures (cells and organs) and external structures (what the organism looks like) of living things to discover how organisms are similar and how they are different. They use microscopes and other forms of technology to compare organisms in a very detailed way (**Figure 3**). For example, they can compare the cell structure of different organisms. They can also compare organisms from around the world to discover how different organisms may be related.

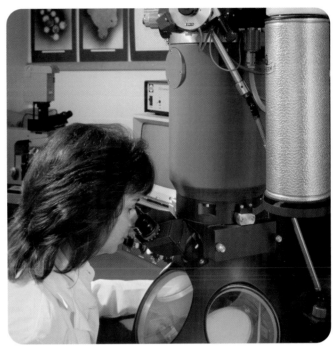

Figure 3
A scientist uses an electron microscope to look at an organism.

⫸ CHECK YOUR UNDERSTANDING

1. Copy the following chart into your notebook. List at least five classification systems that you use to organize things. Explain how each classification system makes your life simpler. The first row is filled in for you.

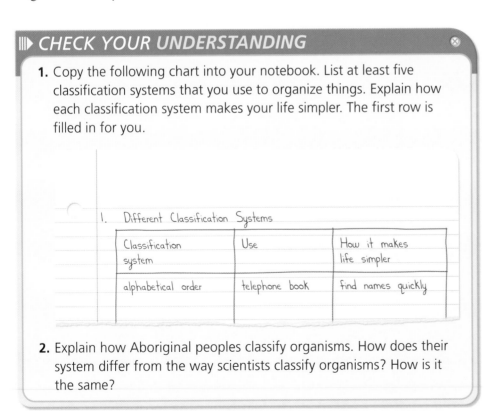

1. Different Classification Systems

Classification system	Use	How it makes life simpler
alphabetical order	telephone book	find names quickly

2. Explain how Aboriginal peoples classify organisms. How does their system differ from the way scientists classify organisms? How is it the same?

2.2 *The Key to Classification*

Scientists classify organisms by looking at their characteristics. They then develop questions that will help them identify and classify organisms. For example, they might ask whether an organism makes its own food. If the answer is yes, then the organism is probably a plant. If the answer is no, the organism is not a plant. This set of questions is called a classification key. It helps scientists find similarities and differences among a group of organisms. Let's see how a classification key works.

Identifying Berries Using a Key

Aboriginal peoples have learned to recognize edible berries and to avoid poisonous berries and berries that taste bad. They carefully observe the leaves of a plant and the way that the berries grow on the plant. How would you know which berries were safe to eat in the woods?

Which of the berries shown in **Figure 1** would be good to eat? You can use a classification key to help you decide. Look at the key in **Figure 2**. What characteristics are being described?

▷ **LEARNING TIP**

If you want to know more about how Aboriginal peoples classify things, ask your teacher if it is possible to bring in an Aboriginal Elder as a guest speaker.

Figure 1
Which of these berries would you eat?

1. a	berries grow along the stem		go to 2
1. b	berries grow in one area on the stem		go to 3
2. a	single berries		huckleberries = edible
2. b	berries in clusters on a stem		go to 3
3. a	single leaves with teeth	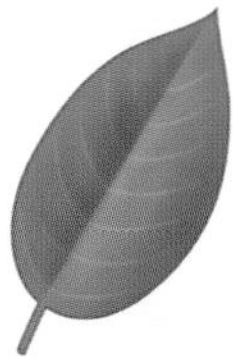	wild cherry = edible but can taste bitter
3. b	leaves divided into leaflets from a central stalk		go to 4
4. a	velvet red berries		sumac = edible but tastes sour
4. b	dark blue berries		blue elderberry = edible

To use the classification key, choose one of the berries shown in **Figure 1**. Then read statements 1. a and 1. b in **Figure 2** to see which statement describes the berry. Then follow the directions in the next column. Keep following the key until you find the name of the berry.

Figure 2
A classification key

The key first asks you to look at where the berries are found on the plant (whether they grow along the stem or only on one place on the stem). You then consider whether the berries grow in clusters or as single berries. The next step is to look at the leaves of the plant. Finally, the key uses the colour of the berries. All of these characteristics describe the external, physical structures of the plant.

Use the classification key to name and classify the berries in **Figure 1**. Which berries would you eat?

IIII▶ CHECK YOUR UNDERSTANDING ⊗

1. How would you explain a classification key to someone? Write three or four sentences that explain what a classification key is, how it is used, and when it would be a useful tool.

2. Give two reasons why it is important to be able to identify an organism.

3. Can you name another characteristic of berries that could have been used in the key?

Solve a Problem

How Can Plants be Classified?

▷ **LEARNING TIP**

To review the steps in problem-solving, see the Skills Handbooks section "Solving a Problem."

Problem

A botanist has just moved into your community. **Figure 1** shows some of the plants that she has classified. She wants to classify the local plants and has asked you to help.

Leafy aster

Figure 1
Native British Columbia plants

Deer fern

Western flowering dogwood

Task

Create an effective classification key to identify five plants based on their leaves and stems.

Criteria

To be successful, your classification key must

- use external, physical characteristics of the leaves and stems
- provide two choices for each of the characteristics in your key
- be accurate and reliable

Plan and Test

1. With a partner, collect a leaf that is attached to a stem from five different plants. Look for specimens that are already on the ground.
2. Label each specimen with a name or a number.
3. Select three or four characteristics that you can use to create a classification key for the five leaves (**Figure 2**).
4. Create a classification key.

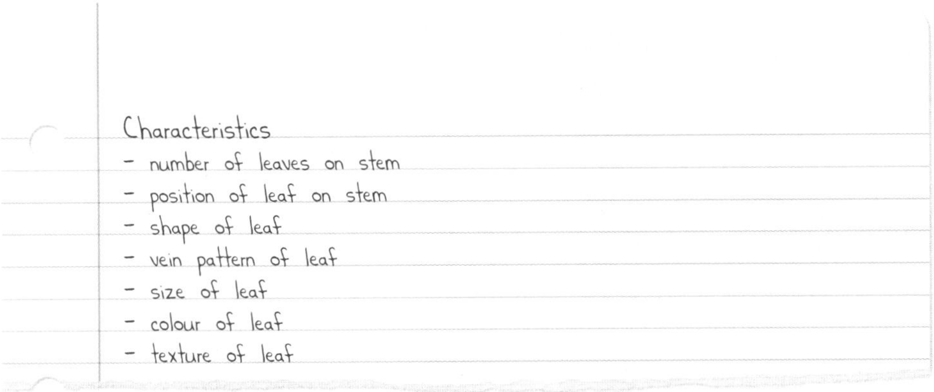
Characteristics
- number of leaves on stem
- position of leaf on stem
- shape of leaf
- vein pattern of leaf
- size of leaf
- colour of leaf
- texture of leaf

Figure 2

Evaluate

5. Explain your classification key to a classmate. Ask for feedback as to how you could make your key clearer.

Communicate

6. Exchange leaves and keys with a classmate. Ask him or her to use your classification key to identify your five leaves. Was your classmate successful?

CHECK YOUR UNDERSTANDING

1. How did feedback from a classmate help you develop a better classification key?
2. Why are observation skills so important when creating and using a classification key?

Healing with Plants

Long before there were doctor's offices and pharmacies, the Aboriginal peoples of British Columbia were experts on using plants to ease pain and treat illnesses.

The Coast Salish peoples, for example, used the leaves of the stinging nettle plant (**Figure 1**) to treat aches and pains. The Interior peoples brewed a tea from the twigs and leaves of the wild lilac plant to ease rheumatism and arthritis pain and to cure diarrhea. The Shuswap peoples had a different purpose for the wild lilac. When left boiling, it was an excellent insect repellent. The Ulkatcho [ul-GAT-cho] (Williams Lake) peoples used the yarrow plant (**Figure 2**) to treat sore muscles. The leaves of the yarrow plant were used as a mosquito repellent by rubbing them on the skin or tossing them in a fire. The roots were used to make a tea that could cure a stomachache.

Today, there are fewer and fewer people in Aboriginal communities who have detailed knowledge of medicinal plants to pass on to younger generations.

Dr. Nancy Turner (**Figure 3**) is working with Aboriginal Elders to preserve this knowledge. For the past 30 years, she has worked closely with Aboriginal Elders in British Columbia to document their knowledge and understanding of plants and ecosystems. Dr. Turner considers the Elders to be teachers and friends. Her hope is that their valuable knowledge will be preserved for the benefit of their communities and the world.

Figure 1
A stinging nettle plant

Figure 2
A yarrow plant

Figure 3
Dr. Turner is an ethnobotanist—a person who studies the classification and uses of plants in different human societies.

How Scientists Classify Living Things

Scientists use classification to help them understand the diversity of life on Earth. They look at the characteristics of living things and then develop questions to ask. For example, they ask whether the organism is made of one or more cells, how it gets the nutrients and energy it needs, and how it reproduces. The answers allow scientists to identify and classify any organism they discover.

One Cell or More

People have always recognized two groups of living things: plants and animals. However, after the development of magnifying tools, such as microscopes, scientists discovered new organisms that were invisible to the naked eye (**Figure 1**). They also began to study the internal structures of organisms. With this knowledge, scientists had a much better picture of how organisms were alike and how they were different.

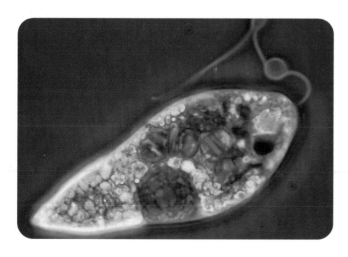

Figure 1
Some organisms, such as *Euglena*, cannot be viewed without a microscope.

> **LEARNING TIP**

The prefix *uni-* means "one" or "single." The prefix *multi-* means "more than one." Think of other words you know that use these prefixes.

Using microscopes, scientists discovered that organisms have very different internal structures. One important difference is whether an organism is a single cell or is made of more than one cell. As you learned in Chapter 1, cells are the basic unit of life—the building blocks that make up all organisms. Scientists discovered that some organisms have a very simple structure. Their entire body is just one cell big. They are **unicellular** [YOO-nee-SELL-yur-luhr], or made up of one cell (**Figure 2**).

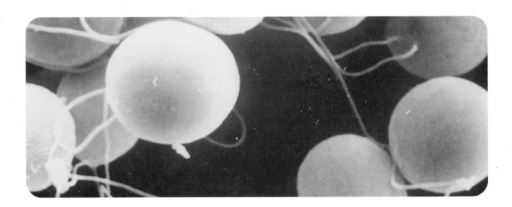

Figure 2
Chlamydomonas is a unicellular green alga.

Other organisms are more complex. They are **multicellular** [MULL-tee-SELL-yur-luhr], or have more than one cell in their bodies (**Figure 3**). In fact, most of these organisms are made up of trillions of cells. Multicellular organisms have different types of cells that perform different functions. Human beings, for example, have bone cells, skin cells, blood cells, and many other types of cells. Each type of cell has its own structure and specific purpose. All the cells of an organism work together to ensure that the organism can perform the functions it needs to survive.

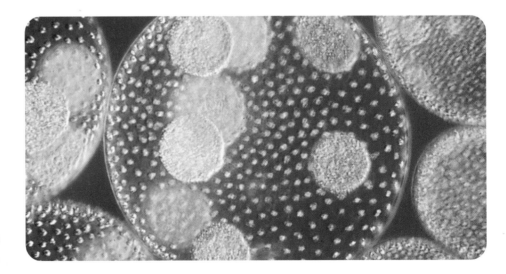

Figure 3
Volvox is a multicellular green alga that is made of thousands of cells.

Plant and Animal Cells

There are many different types of cells. Cells have different sizes, shapes, and functions. Let's look at some of the features of a typical plant and animal cell (**Figure 4**).

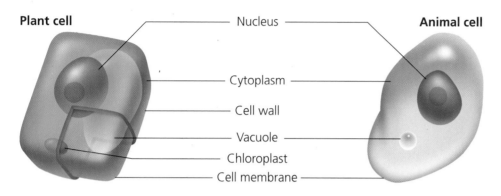

Plant cell

Nucleus

Animal cell

Cytoplasm

Cell wall

Vacuole

Chloroplast

Cell membrane

Figure 4
A plant cell and an animal cell

Both plant and animal cells have a **cell membrane.** The membrane is a thin covering around the entire cell. It encloses the cell's contents. It acts like a gatekeeper by allowing useful materials to move into the cell and waste to move out.

The innermost part of a cell is the **nucleus** [NOO-klee-us]. The nucleus acts as the control centre of the cell. It directs all of the cell's activities, such as movement and growth.

Much of a plant or animal cell is filled with a thick liquid called cytoplasm [SIGH-tuh-pla-zum]. The cytoplasm is where the work of the cell is carried out, as directed by the nucleus. Within the cytoplasm are bubble-like vacuoles [VAK-yoo-ole]. Vacuoles store water and nutrients.

Important differences exist between plant and animal cells. Unlike animal cells, plant cells are enclosed by a cell wall. The cell wall helps to protect the cell and provides support for the plant. Plant cells also contain **chloroplasts** [KLOR-uh-plahst], which are the parts of the cell that contain **chlorophyll** [KLOR-uh-fill]. Chlorophyll gives plants their green colour.

LEARNING TIP

Look at each of the highlighted words on the page. To help remember what they mean, find each word in the diagram. Then read the paragraph that describes the term. Now try to define each term using your own words.

TRY THIS: *MAKE MODELS OF CELLS*

Skills Focus: communicating

Use modelling clay to make a model of a plant cell and a model of an animal cell. Include all of the parts of a cell that were described in this section.

The Five-Kingdom Model of Living Things

As scientists learned more about the internal and external structures of organisms, they discovered that some organisms were like both plants and animals. Others didn't have the characteristics of either plants or animals. Modern scientists realized that there were at least five categories of living things, which they called **kingdoms.** Figure 5 shows the five kingdoms of life. Each kingdom has important characteristics that all of its members have in common.

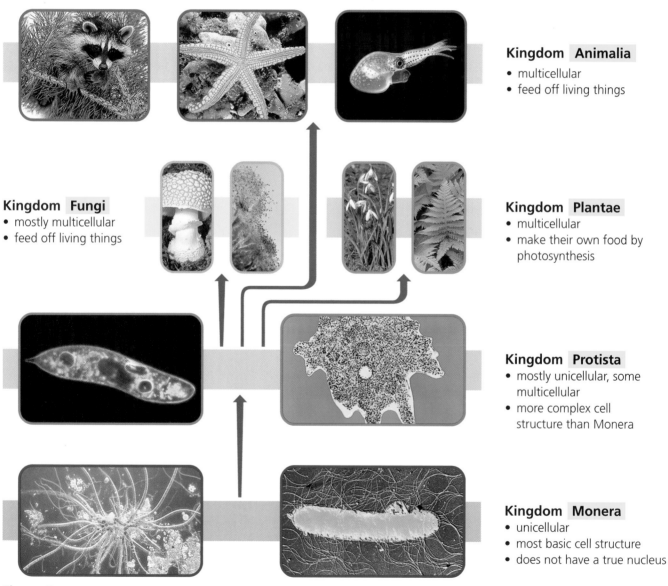

Kingdom Animalia
- multicellular
- feed off living things

Kingdom Fungi
- mostly multicellular
- feed off living things

Kingdom Plantae
- multicellular
- make their own food by photosynthesis

Kingdom Protista
- mostly unicellular, some multicellular
- more complex cell structure than Monera

Kingdom Monera
- unicellular
- most basic cell structure
- does not have a true nucleus

Figure 5
The five kingdoms of life

An International Classification System

Classifying organisms by kingdom is a good beginning. But there are many organisms in each kingdom. For example, there are over one million different types of organisms in the Animal kingdom. Scientists need a way to classify organisms into smaller, more manageable groups. They also need a common way of identifying and naming organisms so that they can describe, compare, and communicate their knowledge about different organisms.

Today, scientists around the world use a single scientific classification system for naming and classifying all organisms. This system was created, in 1735, by Carolus Linnaeus (**Figure 6**). Linnaeus was a Swedish scientist who was very curious about all of the living things he observed. He was the first scientist to divide living things into groups called kingdoms, although he proposed only two kingdoms: plants and animals.

Figure 6
Carolus Linnaeus developed a classification system to organize living things.

Linnaeus proposed that each kingdom could be further classified into a series of smaller categories. His system of classification has seven categories in total. After kingdom comes phylum [FI-luhm], class, order, family, genus [JEE-nuhs], and finally the category of **species.** Organisms that belong to the same species are capable of breeding together and having offspring that can also reproduce.

▷ **LEARNING TIP**

When you read a diagram, make sure to read the caption for help in understanding what the diagram shows. In **Figure 7**, read each sentence in the caption and check that you can see how the information is shown in the diagram.

Figure 7 shows how the seven-category system works. Follow the mountain lion, shown on the far right of the diagram, down through the levels. The mountain lion is a different species than the tiger, even though they are in the same family. So a mountain lion could not reproduce with a tiger.

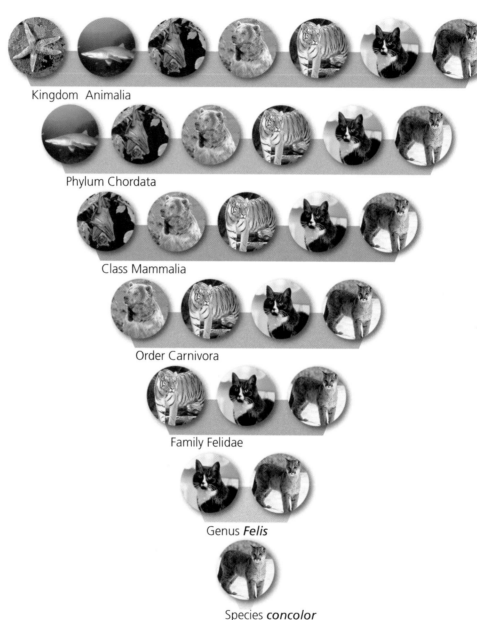

Kingdom Animalia

Phylum Chordata

Class Mammalia

Order Carnivora

Family Felidae

Genus *Felis*

Species *concolor*

Figure 7

The mountain lion is a multicellular organism that gets its food from other living things, so it is a member of Kingdom Animalia. It has a backbone, so it is in the phylum Chordata. It is a mammal, so it belongs in class Mammalia. It feeds on meat, so it is in the order Carnivora. It is a cat, so it is a member of the family Felidae. It is placed in the genus *Felis*, along with the house cats we keep as pets. Finally, it is a member of the species *concolor*, a mountain lion.

A System of Naming

Linnaeus also developed a two-part system for naming organisms. This naming system made it easier for scientists around the world to communicate about the same organism. In this system, every organism has two Latin names. Just as you have a first and last name, each organism has a first name, called the genus name, and a last name, called the species name. The two-part Latin name for the mountain lion shown in **Figure 7** is *Felis concolor*. The two-part Latin name for humans is *Homo sapiens*.

⫸ CHECK YOUR UNDERSTANDING ⊗

1. Are you multicellular or unicellular? Refer to the five-kingdom classification system to explain your answer.

2. Copy the following table into your notebook. Complete the table using information from this section.

2. Structures in Cells

Cell part	Found in plant or animal cells, or both	Function
cell membrane		
cell wall		
nucleus		
cytoplasm		
vacuoles		
chloroplasts		

3. Explain why scientists need at least five kingdoms to group living things.

4. Use what you've learned to classify each of the following organisms into one of the five kingdoms.
   (a) This organism has four sharp claws so that it can climb trees and capture its prey.
   (b) This organism is smaller than the tip of a pin, but its cells can do everything it needs to stay alive.
   (c) This organism cannot move around by itself, but it can make its own food.
   (d) This organism attaches to and feeds off other organisms.
   (e) This organism is only one cell, but it is just as important as other living things.

5. A coyote's two-part name is *Canis latrans*. A dog's name is *Canis familiaris*. Are these two animals closely related? Explain your thinking.

A Closer Look at the Animal Kingdom

TRY THIS: LOOK AT DIFFERENT ANIMALS

Skills Focus: observing, inferring, classifying

The Animal kingdom is made up of many different organisms. For example, both a spider and an ant belong to the Animal kingdom. Work with a partner to make two specimen boxes that you can use to observe these small animals. Using a hand trowel or a spoon, carefully lift a spider into one of your boxes and an ant into the other. Observe the spider and the ant. When you have finished observing them, gently place them back where you found them.

1. How are these two animals the same? How are they different?

2. Compare the ant and the spider with another animal, such as a dog or a cat. What are some of the similarities and differences? Why do you think all these organisms belong in the Animal kingdom?

There are more than one million different species in the Animal kingdom. All animals are multicellular organisms that get their nutrients and energy by eating other organisms. But animals come in a great variety of forms, from spiders to sparrows to sponges (**Figure 1**). To better understand the diversity of animal life, scientists classify animals into groups based on their internal and external structures.

Figure 1
Although they may look like plants, sponges are simple animals.

Vertebrates and Invertebrates

Scientists divide all the organisms in the Animal kingdom into two main groups: **vertebrates** (animals with backbones) and **invertebrates** (animals without backbones). Vertebrates include birds, fish, and mammals. Invertebrates include insects, worms, squids, sponges, sea anemones, and crabs. Vertebrates are the animals that you're most familiar with, but invertebrates are much more common. Scientists estimate that invertebrates make up more than 95% of all animal species. One group of invertebrates—arthropods—includes all the world's insects, shellfish, and spiders. **Figure 2** shows some of the groups that make up the Animal kingdom.

LEARNING TIP ◄

Check your understanding of vertebrates and invertebrates by describing the difference in your own words.

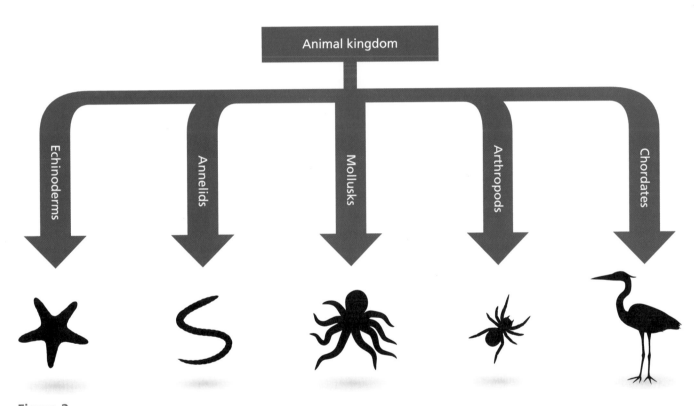

Figure 2
All vertebrates belong to the Chordate group.

Classes of Vertebrates

Scientists divide the vertebrates into classes, based on the internal and external structures they share. There are five main classes of vertebrates: **fish,** **amphibians,** **reptiles,** **birds,** and **mammals.** Each class is defined in **Table 1** on page 36. You are probably familiar with many of the animals in these classes.

Table 1 The Characteristics of Vertebrates

Class	Examples	Characteristics
Fish	salmon, whale shark, ray, seahorse	• live only in water • breathe through gills • use fins to move • lay eggs or give birth to live young • body temperature changes with the environment
Amphibians	frog, toad, salamander	• young live in water and breathe through gills • adults live mainly on land and breathe with lungs • lay eggs in water • young change form as they grow, for example, growing legs • body temperature changes with the environment
Reptiles	crocodile, alligator, lizard, snake, turtle	• some live on land; some live in water • breathe through lungs • most have claws that can be used to dig or climb • many lay soft, leathery eggs on land • body temperature changes with the environment
Birds	eagle, parrot, cardinal, chicken, penguin, puffin	• breathe through lungs • have wings, feathers, and hollow bones • most can fly • lay eggs in a protective shell • hatchlings are cared for by parents • maintain a constant body temperature
Mammals	dolphin, bat, mouse, kangaroo, lemur, human	• most live on land, some live in water • breathe through lungs • most have hair or fur covering their bodies • most give birth to live babies • mother produces milk to feed her babies • maintain a constant body temperature

1. Explain why a spider is considered a member of the Animal kingdom.

2. What characteristic can you use to separate all animals into two groups?

3. Look at **Figure 3**. How can you tell that this animal is not a vertebrate?

Figure 3
An earthworm

4. Use **Table 1** as a guide to help you identify which class of vertebrates an organism would belong to if it had the following characteristics:
   - breathes through lungs and lays eggs that have a shell
   - lives in the water and breathes through lungs (**Figure 4**)
   - lives on land and lays eggs in water
   - has a constant body temperature and gives birth to live babies

Figure 4

2.6 A Closer Look at the Plant Kingdom

Figure 1
Grass and other plants provide food for bison and other living things.

The plant kingdom contains about 300 000 different species. Plants provide the food that all other living things depend on to survive (**Figure 1**). Plants also provide the oxygen we breathe. Plants are found everywhere on Earth, including in water, in soil, and even on rocks.

Like animals, plants are multicellular organisms. But unlike animals, plants make their own food. Plants use energy from the Sun to turn carbon dioxide from the air and water from the soil into food for themselves. They also produce oxygen, which other organisms breathe. Through this process, called photosynthesis, plants grow and become food for other living things (**Figure 2**).

Plants differ from animals in another important way. They are made of plant cells, as you learned earlier in this chapter. Plant cells have a cell wall, which helps to protect the cell and to support the plant. Plant cells also contain chlorophyll, which gives plants their green colour. Plants, unlike most animals, are also stationary and are attached to a surface, like soil.

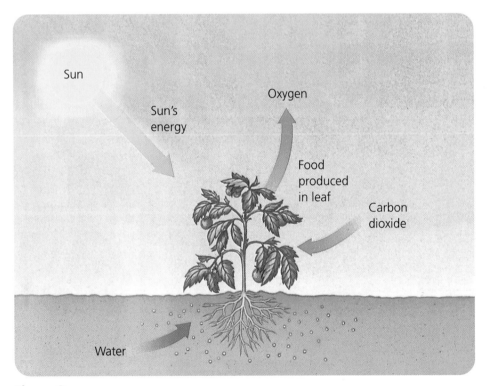

Figure 2
Plants produce their own food and oxygen using energy from sunlight, carbon dioxide from the air, and water from the soil.

Classifying Plants

To classify the thousands of different species of plants, scientists ask three important questions:

- Does it have roots, a stem, and leaves?
- Does it have tube-like structures inside it to help it transport water?
- Does it produce flowers or cones?

On the basis of these three questions, scientists classify plants into four main groups: mosses, hornworts, and liverworts (**Figure 3**); ferns and their relatives; conifers (**Figure 4**); and flowering plants. You are probably most familiar with flowering plants, ferns, and conifers. These three groups of plants have stems, leaves, and roots through which they transport food and water. They also have tubes inside that allow water and food to travel throughout the plant.

Figure 3
Liverworts can be found growing on the ground, on rocks, or even on other plants.

Figure 4
Conifers, such as pine trees, have cones.

TRY THIS: *WATER TRANSPORT IN PLANTS*

Skills Focus: questioning, observing, inferring, communicating

Look at **Figure 5**. What do you think would happen if you dipped the bottom of a stalk of celery into a glass of coloured water? Try it to find out. For best results, leave the celery in the coloured water overnight.

1. What did you see?
2. Explain your observations.

Figure 5

Mosses, hornworts, and liverworts do not have true roots, stems, or leaves, so they depend on their surrounding environment for water. Most of these plants live where it is moist. They are covered in leaf-like structures that allow the plants to absorb the water they require directly into their cells.

TRY THIS: LOOK AT PLANT CHARACTERISTICS

Skills Focus: observing, inferring

Look at a piece of moss and a dandelion plant. Lay the two plants next to each other on your desk so that you can compare them. Use small scissors and tweezers to carefully open the stem or stalk of each plant. Carefully make a cross-section cut in the root of the dandelion plant. Draw a picture of what you see.

Handle scissors and tweezers with extra care.

1. What parts of each plant can you identify? Use **Figure 6** to help you label your drawing.

2. What function does each part play? How does it help the plant live?

3. How are the stem and the stalk the same? How are the stem and the stalk different?

4. How do you think each plant reproduces?

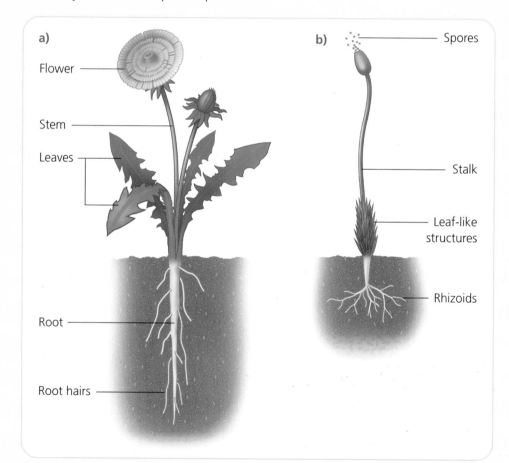

Figure 6
a) A dandelion plant
b) A piece of moss

How Plants Reproduce

Plants are commonly divided into three groups, depending on how they reproduce:

- plants that reproduce with spores
- plants that reproduce with seeds from cones
- plants that reproduce with seeds from flowers

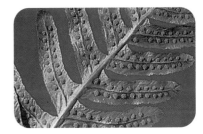

Figure 7
The spores are found under the leaves of a fern plant.

Ferns reproduce by spores. The spores are found on the underside of a fern frond (**Figure 7**). They look like small tufts of soft fluff. The spores are scattered by the wind, fall to the ground, and sprout.

Conifers reproduce with cones. The seeds of the plant are inside these cones. When the cones open, the seeds are scattered by the wind or by animals.

Flowering plants produce seeds in a flower or a fruit. Flowering plants include most trees, shrubs, vines, and flowers. Most fruits and vegetables are also flowering plants.

Look at **Figure 8**. How do you think the seeds are dispersed? Plants that produce fruit have seeds inside the fruit. An animal eats the fruit and scatters the seeds far away. Some plants, such as milkweed, have seed cases that split open and release the seeds. Some plants produce seeds that appear to have wings or parachutes. These structures help the seeds scatter in the wind. Some seeds have tiny hooks that cling to the fur of animals. The seeds are dispersed as the animals move from place to place.

Figure 8

TRY THIS: LOOK AT HOW SEEDS DISPERSE

Skills Focus: observing, inferring

Think about all the ways that seeds can be dispersed. You may want to look in books or survey your neighbourhood. Make a chart to summarize the different methods of seed dispersal. For each method, draw the type of plant and the seed.

▶ CHECK YOUR UNDERSTANDING

1. Discuss why plants are so important to life on Earth.
2. How are plants classified? Into which group would you put the rose?
3. Describe how plants reproduce.

2.7 A Closer Look at the Other Kingdoms

LEARNING TIP

The prefix *micro-* comes from Greek and means "small" or "tiny." So a micro-organism is just a very small living thing. *Micro-* starts many words in science and technology, such as "*microscope*," "*microprocessor*," and "*microwave*." Can you think of other words that start with *micro-*?

Scientists classify the organisms on Earth that are not plants or animals into three kingdoms of life. These kingdoms are: Kingdom Monera, Kingdom Protista, and Kingdom Fungi. The organisms in these kingdoms are not as familiar to us as plants and animals, but they are important to the lives of other living things, including humans.

Many of the organisms in these kingdoms are so tiny they can only be seen with a microscope. They are called **micro-organisms,** which means "tiny life." Micro-organisms can do everything that other organisms can do. They are made of cells, grow and develop, reproduce, and respond to their environment.

Let's take a look at some of the organisms and micro-organisms in the Kingdom Monera, Kingdom Protista, and Kingdom Fungi.

Kingdom Monera

Monerans are the simplest and smallest living things on Earth, but they are also the most widespread. They are unicellular, and do not have a true nucleus. Monerans live only in moist environments. Some Monerans can survive in extremely hot or salty environments, and some can even survive without oxygen! Monerans may also have been the first organisms on Earth.

Kingdom Monera includes one of the most important groups of micro-organisms: bacteria. Bacteria are the most plentiful organisms on Earth. They are present everywhere. **Figure 1** shows the three different shapes of bacteria.

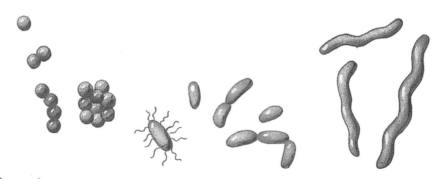

Figure 1
Bacteria come in three shapes: round, rod, and spiral.

Kingdom Protista

Organisms in the Kingdom Protista are more complex than monerans, but most are still only one cell. Protists do, however, contain a true nucleus. This kingdom also includes the simplest multicellular organisms.

Some protists, like algae [AL-jee] and diatoms [DI-uh-toms] (**Figure 2**) are like plants. They contain chlorophyll and they can make their own food through photosynthesis.

Figure 2
There are different types of diatoms. Diatoms make their own food.

Other protists, like paramecia [PAIR-uh-MEE-see-uh] and amoebas [uh-MEE-buhs] (**Figure 3**), are like animals. These organisms are called protozoa [PRO-tuh-ZO-uh]. They feed on other organisms.

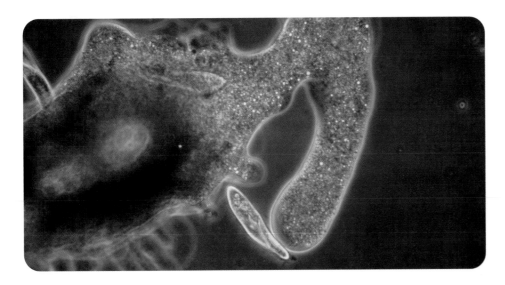

Figure 3
An amoeba engulfs its food.

Kingdom Fungi

Kingdom Fungi includes mushrooms, moulds, and yeasts. They can be unicellular or multicellular. Members of this kingdom are like plants in certain ways—fungi have a cell wall and are stationary. However, they do not make their own food, as plants do. Instead, they grow on or in their food.

Fungi range in size from large mushrooms and toadstools to micro-organisms like yeast (**Figure 4**). Different types of yeast are found in many natural habitats. They are common on the leaves and flowers of plants, both in water and on land. Yeast can also be found living on and inside the bodies of many other organisms, including you!

Figure 4
Yeast is a unicellular organism.

▷ **LEARNING TIP**

It is easier to remember information when you connect it to your own life. What are some examples of harmful and helpful micro-organisms from your life?

What Do Micro-organisms Do?

Some people think that micro-organisms are our enemies. Although some micro-organisms are harmful, most are helpful (**Figure 5**). Life would be very different without them. Micro-organisms are everywhere—in the air we breathe, in the soil around us, in the food we eat, and even in our bodies.

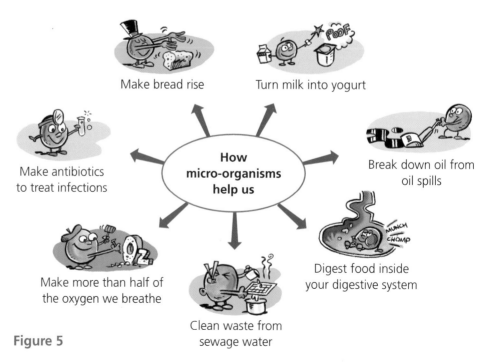

Make bread rise

Turn milk into yogurt

Make antibiotics to treat infections

How micro-organisms help us

Break down oil from oil spills

Make more than half of the oxygen we breathe

Clean waste from sewage water

Digest food inside your digestive system

Figure 5

Unfortunately, not all micro-organisms are helpful. They can spoil food by covering it in mould. Some bacteria, such as *Escherichia coli (E. coli)*, can make you very sick. *E. coli* can live in meat products and is killed only if the meat is cooked thoroughly. Other micro-organisms cause diseases such as whooping cough, tetanus, and malaria. Malaria is caused by a micro-organism from the Kingdom Protista. This disease is transmitted when an infected mosquito bites a person.

⫸ CHECK YOUR UNDERSTANDING ⊗

1. Copy the following table into your notebook. Use information from this section to complete the table.

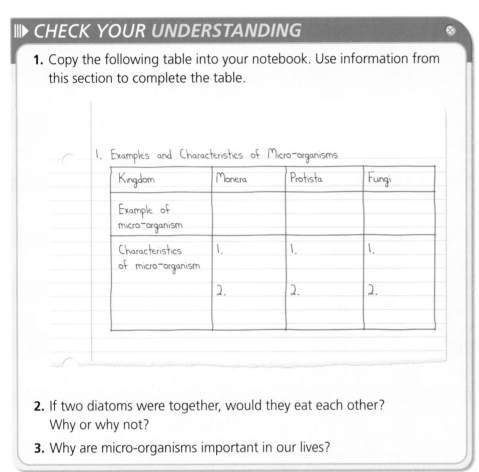

1. Examples and Characteristics of Micro-organisms

Kingdom	Monera	Protista	Fungi
Example of micro-organism			
Characteristics of micro-organism	1. 2.	1. 2.	1. 2.

2. If two diatoms were together, would they eat each other? Why or why not?
3. Why are micro-organisms important in our lives?

2.7 A Closer Look at the Other Kingdoms

2 Chapter Review

Classifying living things helps us understand the diversity of life.

Key Idea: People use classification systems to organize the diversity of living things.

Vocabulary
classification system p. 19

Key Idea: Living things can be unicellular or multicellular.

Vocabulary
unicellular p. 28
multicellular p. 28

Key Idea: Scientists classify organisms into groups based on internal and external features.

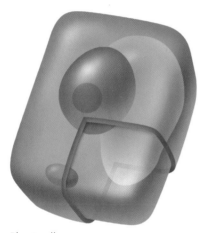
Plant cell

Animal cell

Vocabulary
cell membrane p. 29
nucleus p. 29
chloroplasts p. 29
chlorophyll p. 29
vertebrates p. 35
invertebrates p. 35
fish p. 35
amphibians p. 35
reptiles p. 35
birds p. 35
mammals p. 35
micro-organisms p. 42

Key Idea: Scientists classify living things into five kingdoms: Animalia, Plantae, Fungi, Protista, and Monera.

Kingdom Animalia   Kingdom Plantae   Kingdom Fungi

Kingdom Protista   Kingdom Monera

Vocabulary
kingdoms p. 30
Animalia p. 30
Plantae p. 30
Fungi p. 30
Protista p. 30
Monera p. 30
species p. 31

Review Key Ideas and Vocabulary

When answering the questions, remember to use the chapter vocabulary.

1. Give two examples of how people use classification systems to help them organize their knowledge about living things.

2. Give an example of an internal structure and an example of an external structure that scientists use to classify organisms.

3. Every organism is either a single cell or made of more than one cell. What are the names for these two groups of organisms?

4. Name the five kingdoms of living things.

Use What You've Learned

5. Look carefully at **Figure 1**. List three characteristics that you could use to make a classification key for these organisms.

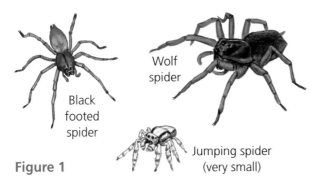

Wolf spider

Black footed spider

Jumping spider (very small)

Figure 1

6. Look at the organisms in **Figure 2**. Explain how the organisms are alike. Explain how they are different.

Bracket fungi   Bindweed

Figure 2

7. If you were given an organism that looked like a plant, what characteristics would you look for to find out if it was actually a plant?

8. Imagine that you have discovered a new organism. Draw a picture of the organism. Which kingdom do you think it belongs in? Explain why.

9. Explain how each of the seeds in **Figure 3** are dispersed.

Figure 3

Think Critically

10. How can you distinguish between two species that look very similar? The two owls in **Figure 4** are different species. What external characteristics might a scientist have used to decide that they were not the same species?

Strix occidentalis (spotted owl)

Strix varia (barred owl)

Figure 4

11. Could the five-kingdom classification system change as scientists discover new organisms? Explain your thinking.

Reflect on Your Learning

12. What is the most important idea that you have learned about diversity? Explain why you think this idea is important.

CHAPTER 3

Magnifying tools make the invisible world visible.

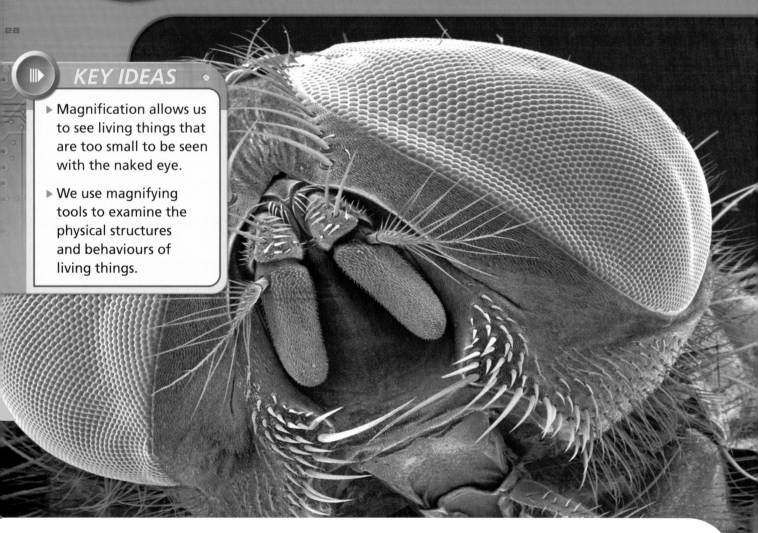

KEY IDEAS

▶ Magnification allows us to see living things that are too small to be seen with the naked eye.

▶ We use magnifying tools to examine the physical structures and behaviours of living things.

▷ **LEARNING TIP**

Before you begin this chapter, take a quick look through, noticing the headings and subheadings, photos, and activities. Predict what you will learn in this chapter.

If you could shrink yourself to the size of an ant, the world would look very different. You would be able to observe small things up close. You would be able to see the compound eyes of a fly, as shown in the photo, and many other wonders of nature!

Even though we can't shrink to the size of an ant, we have tools, such as magnifying lenses and microscopes, that help us see tiny things up close. Microscopes make the invisible world visible to us. And what a fascinating world it is! Using different magnifying tools, scientists have discovered an ever-expanding world of tiny creatures living in us, on us, and around us.

Discovering Life Up Close

TRY THIS: *LOOK AT A LEAF*

Skills Focus: observing, inferring

Make a tracing of a leaf on graph paper. Then examine the leaf carefully. Add all the details you can see to your tracing. Describe both your qualitative observations (texture and colour) and quantitative observations (length of leaf and number of lobes). Now use a magnifying glass to examine the leaf. Add these details to a section of your drawing. What details can you see that you were not able to see using just your eyes?

Have you ever noticed that when you look closely at something, such as a leaf, you see all kinds of details that you don't see when you look at the object from far away? But even when you look at something very closely, there are limits to what you can see. To see more, we need to **magnify,** or make the object look larger than it really is. Magnifying tools allow us to see very small organisms, like those in the Kingdom Protista and the Kingdom Monera, and also to see structures, such as cells, in larger organisms.

Look at **Figure 1**. The children in the photo are using a magnifying glass to look at objects on a log, just as you used a magnifying glass to look at a leaf in the Try This activity. A magnifying glass is made of one lens in a frame with a handle. A lens is a curved piece of glass that magnifies things so that they appear larger.

Figure 1
Looking through a magnifying glass makes small objects appear larger.

The Romans used magnifying glasses almost 2000 years ago, but different forms of magnifiers were used much earlier. Early magnifiers were given very interesting names. Their names showed how they were used. For example, "reading stones" were laid on top of print to magnify the letters and "flea glasses" were used to study tiny organisms.

Skills Focus: observing, inferring

Make your own magnifying tool. Place two pencils on a printed page, about 2 cm apart. Put a small piece of clear tape across the pencils. Look at the letters through the tape. Then put a drop of water on the tape. Look at the letters on the page through the drop of water (**Figure 2**).

1. Compare what you saw when you looked at the letters through only the tape with what you saw through the tape with the drop of water.

2. What does the drop of water do? What can you call the drop of water?

3. What do you think would have happened if you had used a bigger drop of water? Try it and see. Which size of drop made the best magnifier?

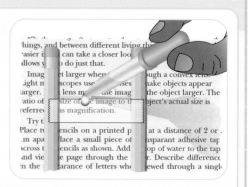

Figure 2

The Development of Modern Microscopes

Using a magnifying glass allowed people to take a closer look at objects. But it wasn't until microscopes were developed that people could see the details in a drop of blood or in a butterfly's wing. A **microscope** is a device that uses a lens or a system of lenses to greatly magnify the image of an object.

The earliest microscopes were made by Anton van Leeuwenhoek in the 1660s. Leeuwenhoek was able to grind and polish glass lenses to magnify objects up to 200 times their real size. Leeuwenhoek's microscopes used one lens to magnify objects. This type of microscope is called a simple microscope. Using his microscopes, Leeuwenhoek saw what he called "animalcules" in pond water (**Figure 3**). These organisms are now known as protozoa and bacteria, which belong to the Kingdom Protista and the Kingdom Monera.

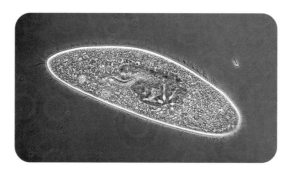

Figure 3
Leeuwenhoek would have seen this protozoan, called a paramecium, when he looked through his microscopes.

The microscopes that we use today have two lenses. Two lenses give much greater magnification than a single lens. Try looking at this page using two magnifiers. A microscope with two lenses is called a compound microscope. Another microscope that is used today is an electron microscope. With an electron microscope, scientists can see much smaller things (**Figure 4**).

Figure 4
A flea and a flea's leg viewed with an electron microscope. An electron microscope is able to achieve magnification of 1 000 000×. This means that an object will look 1 000 000 times larger than it is.

Microscopes are used in many fields. Scientists use microscopes to study organisms. Police investigators use microscopic evidence to solve crimes. Doctors use microscope technology to perform surgeries. Microbiologists use microscopes to identify germs and diseases. Microscopes have greatly increased our knowledge of living things.

▐▶ CHECK YOUR UNDERSTANDING ⊗

1. How does a drop of water act like a magnifier?
2. Why do people continue to work on new technologies for magnifying things?
3. What did Anton van Leeuwenhoek discover with his microscopes? Do you think he would have made this discovery without a microscope? Explain.
4. What is the difference between a compound microscope and the simple microscopes that Leeuwenhoek used?
5. List three ways that microscopes are used today.

LEARNING TIP ◁

Do not guess when you are answering questions. Look back through the section to find the answers. Even if you think you know the answer, it is always a good idea to go back and check the text.

3.2 Learning How to Use a Microscope

The compound microscope uses two lenses—the ocular lens and the objective lens—to magnify an object. **Figure 1** shows the parts of a microscope and explains what they do.

Ocular lens: This is the lens that you look through. It is also called the eyepiece. The ocular lens usually magnifies the image produced by the objective lens by 10×.

Revolving nosepiece: This rotates so that the objective lenses can be changed.

Objective lenses: These three lenses are just above the object, or specimen, that you view. The low-power lens magnifies the specimen about four times (4×). The medium-power lens magnifies by ten times (10×), and the high-power lens magnifies by forty times (40×).

Stage: This platform has an opening in it just under the objective lenses. You place your slide on the stage, with the object you want to view above the opening. Clips hold the slide in place.

Diaphragm: This controls the amount of light that reaches the object you are viewing.

Coarse-adjustment knob: This is used to move the object into focus. It is only used with the low-power lens.

Fine-adjustment knob: This is used only with the medium-power and high-power lenses. It brings the object into sharp focus. You use it only after you have located the object under low magnification using the coarse-adjustment knob.

Light source: You need light below the object to see it clearly. Some microscopes use a mirror as the light source.

Figure 1
The parts of a microscope

Preparing a Slide of Onion Skin

When you want to view an object under a microscope, you place it on a thin rectangular piece of glass called a **slide.** The object on the slide is then covered with a thin glass square called a **cover slip.** A glass slide and cover slip are shown in **Figure 2**.

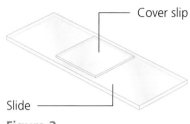

Figure 2

Preparing a slide takes practice. You must learn how to make a specimen thin, so that light can shine through the slide. You must also learn how to place a cover slip over a specimen. Follow these steps to make a slide of a piece of onion skin.

1. Use tweezers to remove the thin skin, or membrane, from the inside of a layer of an onion.

2. Place a small piece of onion skin on a slide. Try not to trap any air bubbles under the skin.

 Be careful when handling glass slides and cover slips. The edges of the slides can be rough.

3. Use an eyedropper to place a drop of water on the onion skin.

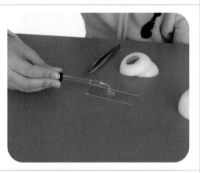

4. Hold a cover slip between your thumb and forefinger. Place the edge of the cover slip on one side of the skin at a 45° angle. Carefully lower the cover slip to cover the skin. Gently tap the slide with the eraser end of a pencil to remove any air bubbles.

Focusing a Microscope

Every time you look at a slide under a microscope, you need to focus the microscope twice: once with the coarse-adjustment knob and once with the fine-adjustment knob. Follow these steps to learn how to focus, so that you can view an object clearly.

1. Turn the nosepiece so that you are looking through the low-power objective lens.

2. For your safety, pick up your slide by the edges using your thumb and forefinger. This will also ensure that nothing touches your specimen. Place your slide on the stage. Make sure that the specimen is above the opening. Move the clips over the slide to secure it in place.

Watch the stage as you use the coarse-adjustment knob so that the lens doesn't break the slide and get damaged.

3. Turn the coarse-adjustment knob slowly until the slide is in focus. Be careful that the lens does not touch the slide. You may need to adjust the diaphragm to increase or decrease the light. What can you see?

4. Turn the nosepiece to the medium-power lens.

5. Use the fine-adjustment knob to bring the slide into focus. How has the image changed?

 Never use the coarse-adjustment knob with the high-power lens as it can break the cover slip or damage the lens.

TRY THIS: *TAKE A CLOSER LOOK*

Skills Focus: observing

Now that you have learned how to use a microscope, you can take a closer look at some of the things around you. Prepare slides of salt, chalk dust, a leaf, and a thin slice of cork, or anything else you think would be interesting. Look at the slides under a microscope. What do you see?

▶ CHECK YOUR UNDERSTANDING

1. Which part(s) of a microscope perform each job described below?
   - holding the slide in place
   - magnifying the specimen you are viewing
   - turning so that you can use the lens you want
   - focusing the specimen

2. Why would you not want air bubbles trapped under or over a specimen?

3. Why should you not allow the lens to touch the slide?

SKILLS MENU

○ Questioning	● Observing
○ Predicting	○ Measuring
○ Hypothesizing	● Classifying
○ Designing Experiments	● Inferring
○ Controlling Variables	○ Interpreting Data
○ Creating Models	● Communicating

Observing Pond Water

Many tiny organisms live in pond water. Using a magnifying glass and a microscope, you will see tiny single-celled protists that are plant-like because they make their own food. You will also see some single-celled protists that are animal-like because they hunt and gather other organisms for food. You will also see algae, which look like long chains of cells that contain chloroplasts. Algae belong to the Kingdom Protista. As well, you will see small organisms, such as water fleas, insect larvae, copepods, and hydra that belong to the Kingdom Animalia. **Figure 1**, on the next page, shows some of the organisms you may see in pond water.

You will see that these organisms move in many ways. Some micro-organisms, such as *Euglena* and *Volvox*, move by whipping a tail called a flagellum [fluh-JELL-um]. Paramecia are covered with tiny hairs, called cilia [sill-EE-uh], that wave back and forth to move. Other pond organisms, such as the hydra, can glide or somersault along using tentacles. Water fleas appear to hop along.

In this investigation, you will observe organisms in pond water, using different degrees of magnification. You will also observe the movements and feeding behavior of these organisms. Use **Figure 1** to identify some of the organisms you see.

Question
What organisms can you observe in pond water?

apron

pond water

magnifying glass

petri dish

eyedropper

cover slip

slide

microscope

paper towels

Materials
- apron
- pond water in white container (such as a margarine container)
- magnifying glass
- small petri dish
- eyedropper
- microscope
- slide and cover slip
- paper towels

> ✋ Remember to carry the microscope using both hands, and to adjust the focus and carry the slides carefully.

Hydra

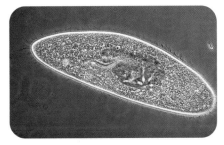

Paramecium

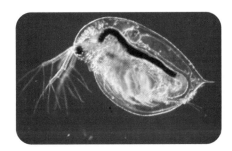

Water flea

Amoeba

Copepod

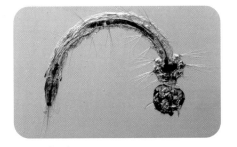

Mosquito larva

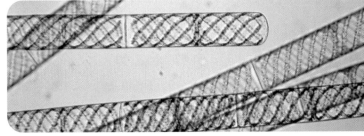

Spirogyra

Figure 1

Procedure

1 Copy the following table in your notebook. Record all your observations in your table.

Observations				
Method of observing (naked eye, magnifying glass, microscope)	Drawing of organism and name	Description of structure	Movement	Feeding behaviour

2 Look at the container of pond water. Draw all the organisms that you see and, if possible, write their names in your table. Record details about their structure, movement, and feeding behaviour.

3 Use an eyedropper to pick up any organisms that you cannot see clearly and transfer the organisms to a small petri dish. Put the petri dish on a piece of white paper. Use a magnifying glass to make your observations. Record your observations in your table.

3.3 Conduct an Investigation

4 Use the eyedropper to place one drop of pond water on the centre of a slide. Touch a cover slip to the slide at a 45° angle. Gently lower the cover slip, being careful not to trap any air bubbles.

 Use only the fine-adjustment knob when you focus the high-power lens so you do not break the cover slip.

5 Place the slide on the stage of a microscope. Look for organisms using the low-power lens. If you find organisms that are too small to see clearly, try looking at them using the medium-power lens. The high-power lens is usually not useful for observing pond water organisms, but you can try it.

6 Draw all the organisms you see and write the observations in your table.

7 Wipe your slide clean with a paper towel. Repeat steps 5 and 6 to look for other organisms.

8 Dispose of your slide as directed by your teacher. Wash your hands.

Analyze and Evaluate

1. How many organisms were you able to identify using only your eyes? Using a magnifying glass? Using the low-power lens of a microscope? Using the medium-power lens? Did you see any organisms using the high-power lens?

2. How did the magnifying glass and microscope help you with your observations?

3. What do you think you would be able to see if you used a more powerful microscope?

Apply and Extend

4. Medical lab technicians use microscopes to observe blood samples. Why would a technician need to use a microscope with a high magnification?

5. Can you think of other jobs for which a microscope would be a valuable tool? Explain your thinking.

6. How has the microscope increased our understanding of the diversity of life that exists on Earth?

▶ CHECK YOUR UNDERSTANDING

1. In this investigation, you had to be careful when recording your observations. How could your conclusions about magnification be affected if your observations were not accurate and detailed?

SKILLS MENU

○ Questioning	● Observing
○ Predicting	○ Measuring
○ Hypothesizing	○ Classifying
○ Designing Experiments	● Inferring
○ Controlling Variables	○ Interpreting Data
○ Creating Models	● Communicating

Learning about Water Bears

Canada is home to polar bears, black bears, and grizzly bears. But did you know that it is also home to the "bear" shown in **Figure 1**? The scientific name for these organisms is *tardigrades*, but scientists affectionately call them "water bears."

Water bears are tiny multicellular organisms that belong to the Kingdom Animalia. Their favourite home is a moist environment, preferably a clump of moss. They grow to approximately 0.3 mm in length (about the width of a hair) and move along on eight legs. These amazing creatures can survive in harsh conditions. They have been found under ice and in hot springs. In fact, they have been found everywhere on our planet where there is water.

In this investigation, you will use a microscope to study the physical appearance and behaviour of water bears.

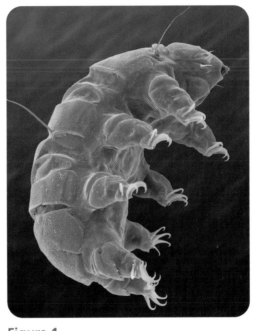

Figure 1
A water bear viewed under an electron microscope.

apron

moss   water

dish   petri dish

cover slip

eyedropper   slide

petroleum jelly

microscope

Question

What are the structural and behavioural characteristics of water bears?

Materials

- apron
- moss
- pond water or rainwater
- shallow dish
- small petri dish
- eyedropper or pipette
- slide and cover slip
- petroleum jelly
- microscope

For a review in using a microscope and preparing slides, see Section 3.2, pages 52–55.

Remember to carry the microscope using both hands. Adjust the focus and carry the slides carefully.

⫸ Procedure

1 Obtain some moss from your teacher. Look at the moss for evidence of living organisms. Sketch the moss and what you see in it.

2 Place the clump of moss in a shallow dish. Pour enough pond water or rainwater into the dish to cover the moss by 1 cm. (Do not use tap water.)

3 Let the moss soak overnight. The next day, take the moss out of the water. Pour out the water left in the shallow dish.

4 Squeeze the moss over a small petri dish so that any water in the moss collects in the petri dish. Shake any extra water out of the moss. Look for evidence of living organisms in the water, and sketch what you see.

5 Examine the water in the petri dish under a microscope, using the lowest power. You should be able to see the water bears using this power. Record your observations.

6 Use an eyedropper or pipette to transfer your water bears to a microscope slide for viewing. To avoid crushing your water bears, place a dab of petroleum jelly on the corners of the cover slip before lowering it onto the slide.

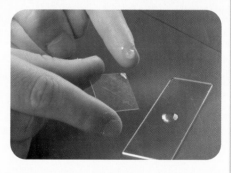

7 Look at the slide under the microscope. Record what you see.

8 Dispose of the slide as directed by your teacher. Wash your hands.

Analyze and Evaluate

1. When you looked at the moss without magnification, what life did you observe?

2. When you looked at the petri dish without magnification, what life did you observe?

3. When you looked at the petri dish using the microscope, what organisms could you see?

4. Could you see any physical structures of the water bear when you used the microscope? How do you think these structures help the water bear survive?

5. What behaviours of the water bear did you observe? How do you think these behaviours help the water bear survive?

6. What did you learn about water bears by examining them under the microscope?

Apply and Extend

7. How does your drawing of a water bear compare with the photo at the beginning of this investigation? Are there differences? If so, can you explain why?

8. Water bears can survive in very harsh conditions. Why do you think scientists might study water bears?

> **▶ CHECK YOUR UNDERSTANDING** ⊗
>
> 1. Why is it important to read and follow the procedure for an investigation carefully?
>
> 2. Identify two steps in this procedure that were important for helping you see the water bears.

Microscope Detectives

Forensic scientists find and analyze clues at crime scenes. Criminals usually leave behind some clues about their identity that a forensic scientist can detect using microscopes. Let's look at some of the microscopic clues these scientists might examine.

Fibres

Do you think that the fibres of one red sweater are the same as the fibres of another red sweater? Look at the fibres in **Figure 1**. Do they look the same? A forensic scientist can analyze fibres found at a crime scene and match them to clothing worn by a suspect.

Gather fibres from several different fabric samples. Look at the fibres under a microscope. How are they different? Simulate wear by rubbing each fibre between your fingers. Look at it again under the microscope to see if there is any difference. Rip each fibre into two pieces and look at the edges. What do you notice?

Figure 1

A fibre found at a crime scene came from the sweater of one of these three suspects.

Fingerprints

No one has exactly the same fingerprints as you. You will grow older and bigger, but the pattern of your prints remains the same throughout your life. Each fingerprint has its own pattern of whorls, arches, and loops (**Figure 2**).

Whorls

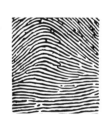

Arches

Loops

Figure 2

Fingerprints have whorls, arches, or loops. Most people's fingerprints have loops.

Forensic scientists carefully examine fingerprints lifted from a crime scene (**Figure 3**). They look at the patterns and the ridges between the patterns.

Use a hand-held magnifier to examine your own fingerprints. Look for whorls, arches, and loops. Compare your fingerprints with a classmate's fingerprints. How are they similar? How are they different?

Figure 3
A scientist compares the fingerprints of the three suspects with the fingerprint found at the scene of the crime.

Pollen and Spore Analysis

Forensic scientists use microscopes to analyze pollen and spores that are found at crime scenes. Much like fingerprints can be used to identify an individual, pollen grains and spores can be used to identify a specific area, or locale. Scientists can use pollen and spore samples to identify where a crime took place. They can also compare pollen samples, taken from a suspect's clothing or shoes, with pollen samples found at a crime scene to link a suspect to the scene of the crime (**Figure 4**).

Analyzing pollen and spores is not commonly used in Canada. However, because pollen and spores are everywhere—from dust to soil to hair—the analysis of pollen and spores may become a widely used tool to solve crimes.

Using microscopes to compare fibres, identify fingerprints, and analyze pollen and spores are just some of the many things that forensic scientists do. Find out more about forensic scientists on the Internet. Share what you learn with your classmates.

www·science·nelson·com

Figure 4
Pollen samples from the three suspects can be compared with pollen found at the crime scene.

Chapter Review

Magnifying tools make the invisible world visible.

Key Idea: **Magnification allows us to see living things that are too small to be seen with the naked eye.**

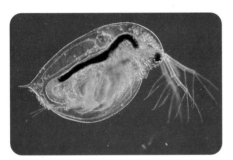

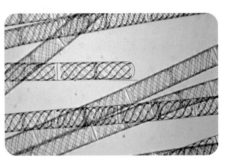

Key Idea: **We use magnifying tools to examine the physical structures and behaviours of living things.**

Vocabulary

microscope p. 50

slide p. 53

cover slip p. 53

Review Key Ideas and Vocabulary

When answering the questions, remember to use the chapter vocabulary.

1. Why are magnifying tools so useful?

2. How do microscopes help scientists understand the organisms on Earth?

Use What You've Learned

3. How is a drop of water similar to a magnifying glass or a microscope?

4. A thick, syrupy substance called methyl cellulose is often used when preparing a slide of moving micro-organisms. Why do you think a biologist would want to watch a micro-organism in this syrupy substance?

5. Write a letter to Anton van Leeuwenhoek explaining how important microscopes are to us today.

6. Create a web, using words and pictures, that shows how microscopes are used today.

7. Create a poster or a comic strip that shows how to use a microscope properly and how to prepare a specimen on a slide.

Think Critically

8. Today, scientists use scanning tunnelling microscopes to look at molecules and large atoms. What kinds of research do you think scientists use these microscopes for? What do you think they might discover?

9. How would our understanding of organisms be different without magnification?

Reflect on Your Learning

10. Suppose that you could look at anything you wished up close. What would you choose? What do you think you would be able to see? How could looking up close change the way you normally think about the object?

CHAPTER 4

Living things adapt to their environments.

KEY IDEAS

- Many organisms have structures that help them adapt to their environments.
- Many organisms have behaviours that help them adapt to their environments.
- Environmental changes threaten some species with extinction.

If you look closely at a yellow flower, you may discover that you are also looking at a yellow crab spider. These tiny spiders live on yellow flowers, where they are invisible both to the birds and wasps that feed on them, and to the insects they rely on for their own food.

All living things have structures and behaviours that allow them to meet their needs and to survive in their environment. Like the crab spider's yellow colour, these adaptations help an organism live long enough to reproduce.

In this chapter, you will look at some of the fascinating ways in which living things are adapted to their environment. You will also look at what happens to organisms that cannot adapt to changes in their environment.

Characteristics for Survival

All organisms have characteristics that help them survive in their environments. These characteristics are called adaptations. Some adaptations are structures. For example, some plants have brightly coloured flowers to attract birds and insects for pollination. Cacti, which grow in dry areas, have fleshy stems to store water and short prickly leaves to reduce water loss.

Some adaptations are behaviours that help organisms survive. **Behaviours** are what organisms do, whether it is swimming, flying, or sleeping. Hibernation is an example of a behaviour that helps some organisms survive cold winter temperatures. The great variety of structures and behaviours of organisms is responsible for the diversity of life on Earth.

LEARNING TIP ◁

Check that you understand the two types of adaptations that help organisms survive by explaining them to a classmate.

TRY THIS: LOOK AT A HUMAN ADAPTATION

Skills Focus: observing, inferring

Take a look at your thumb. It is called an opposable thumb because it can touch all the fingers on the same hand. Your thumb makes it possible for you to do many things that animals without opposable thumbs cannot do.

Have a partner time how long you take to untie one of your shoes, take it off, put it on again, and tie it again. Record your time. Now tape your thumb firmly to the rest of your hand so that you cannot use it. Try the shoe-tying task again. Record how long you take.

1. How useful is having an opposable thumb?

2. Apes, chimpanzees, and other primates (including humans) have opposable thumbs. How is this adaptation useful for helping these animals survive?

Feet for Many Purposes

Animals have feet of many sizes and shapes that are perfect for swimming, perching, climbing, grasping, or walking on mud. These adaptations have allowed the animals to survive in their environments. For example, whales and dolphins have flippers. Ducks and penguins have webbed feet that are great for swimming and for walking in muddy areas or snow without sinking. Sea otters also have webbed feet to help them swim quickly through the water (**Figure 1**).

Figure 1
Sea otters use their feet like paddles.

Some feet have special toes. A heron has long, spread-out toes that help it stay on top of mud (**Figure 2**). A thrush has three toes that face forward and one toe that faces backward. This shape allows the thrush to perch safely in trees, even while sleeping! A porcupine has sharp claws on its feet to help it climb.

Figure 2
A heron relies on its feet to keep it from sinking in mud.

What about feet for speed? One of the fastest creatures on Earth is the cheetah (**Figure 3**). How are a cheetah's feet built for speed?

Figure 3
A cheetah can run at speeds of 110 km an hour.

Some Owl Advantages

Owls are adapted to live in a variety of habitats, from the Arctic to the dry regions of southern deserts. There are over 140 species of owls in the world. Owls range in size from the large eagle owl of Eurasia, which grows up to 70 cm in length, to the northern pygmy owl, which is no larger than a sparrow. What adaptations make owls so successful? Let's look at some of the structural adaptations that enable owls to survive in so many different habitats.

Eyesight

Like most birds, owls have very large eyes (**Figure 4**). Unlike other birds, which have one eye on each side of the head, an owl's eyes are at the front. Owls cannot move their eyes. They have to turn their heads to look sideways. Owls can turn their heads almost all the way around to see what is behind them. This adaptation helps to protect owls from possible predators sneaking up on them.

Owls can see well in the daylight, but their nighttime vision is amazing. Most owls are active at night. The pupil in an owl's eye can open very wide, allowing the owl to use all the available light. They can recognize and swoop down on a potential meal in almost complete darkness.

Wings, Feet, and Beaks

Owls have wide wings, powerful feet, and a strong, hooked beak (**Figure 5**). These structures help to make owls very good hunters. Owls also have fine, fringed feathers on the underside of their wings. These feathers help to muffle the sound of the air flowing over their wings, so that owls are almost silent when flying. Consider the advantage that this adaptation gives owls when hunting! This adaptation is not present, however, in the few owl species that hunt during the day.

As an owl sneaks up on an animal, it extends its razor-sharp talons to grip its prey. If the animal is too large to swallow whole, the owl can easily rip the animal into bite-sized pieces with its powerful beak.

Figure 4
Owls, such as this screech owl, have very large eyes on the front of their heads.

Figure 5
The barn owl is an excellent night-time hunter, feeding mostly on rodents.

Colouring

Many species of owls have **colouration** that helps them blend in with their environments. This special colouring is called **camouflage.** For example, the head, wings, and back of a burrowing owl are sandy brown, and its chest is white with large brown speckles (**Figure 6**). This colouring provides excellent camouflage in the dry grassland where the owl lives.

The snowy owl has dappled white colouring—perfect for its snowy surroundings (**Figure 7**). Unfortunately, the colour advantage is lost when summer arrives. As the snow melts in the spring, however, the snowy owl moves to sit on patches of snow or ice. Scientists are unsure whether the snowy owl does this to camouflage itself or whether it is just trying to keep cool.

Figure 6
The burrowing owl blends in with its surroundings.

Figure 7
The colouration of the snowy owl provides camouflage in snow.

Symbiosis: A Behaviour for Survival

Symbiosis [SIM-by-O-sis] is an example of a behaviour that helps some organisms survive. In symbiosis, two organisms live together and help each other. Some birds help to keep other animals clean. For example, the oxpecker feeds on ticks and other insects on a rhinoceros' skin. The oxpecker gets food, and the rhinoceros gets rid of the irritating insects.

Lichens [LIE-kuhns] are organisms that result from the symbiotic relationship between a fungus and a green alga (**Figure 8**). The fungus provides the alga with water, while the plant-like alga provides the fungus with food. This relationship allows both the fungus and the alga to survive in environments where they wouldn't be able to survive alone. You will learn more about other survival behaviours in Section 4.3.

Figure 8
Lichens survive in a wide variety of environments including rocks and tree trunks.

▶ CHECK YOUR UNDERSTANDING ⊗

1. Describe four adaptations that show why the owl is a successful organism.

2. Look at the sketches of feet shown in **Figure 9**. Describe how the structure of each foot would be an advantage in a particular environment.

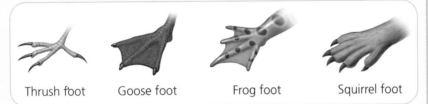

Thrush foot   Goose foot   Frog foot   Squirrel foot

Figure 9

3. Can you spot the fish in **Figure 10**? What adaptation has increased its chances of survival?

Figure 10

4. A cow has billions of micro-organisms in its stomach to help it digest its food. What is this relationship called? How do the micro-organisms help the cow? How does the cow help the micro-organisms?

○ SKILLS MENU

○ Questioning	● Observing
○ Predicting	○ Measuring
○ Hypothesizing	○ Classifying
○ Designing Experiments	● Inferring
○ Controlling Variables	● Interpreting Data
● Creating Models	● Communicating

Examining Bird Beaks

Birds have a variety of different sizes and shapes of beaks to help them get food (**Figure 1**). Some birds use their beaks to crack open seeds, while other birds spear insects. Still other birds use their beaks to tear plants from mud. Their beaks can also strain food from the mud and water. In this investigation, you will examine how bird beaks are adapted to obtain different types of food.

Pelican

Nighthawk

Mallard

Skimmer

Avocet

Crossbill

Warbler

Grosbeak

Swan

Sandpiper

Great blue heron

Falcon

Figure 1

sunflower seeds

tongs

pliers

jellybeans   sprouts

Question

How are bird beaks adapted for birds to obtain food from their environments?

Materials

- sunflower seeds
- jellybeans
- patch of grass or bean sprouts
- needle-nose pliers
- tongs

1 Copy the following table into your notebook. Record all your observations in your table.

Observations for Investigation 4.2

Tool (beak)	Effect on sunflower seed	Effect on jellybean	Effect on grass
tongs			
needle-nose pliers			

2 Pick up a sunflower seed with a pair of tongs. Apply force to the seed. Then pick up a sunflower seed with the pliers and apply force. Record all your observations.

3 Place a sunflower seed in the jaws of the needle-nose pliers in the two positions shown below. Apply pressure until the seed splits or is crushed.

4 Repeat step 3, but this time vary the position of your hand as shown in the photos below. Do not vary the position of the seed. Keep the seed in the position shown in the top photo.

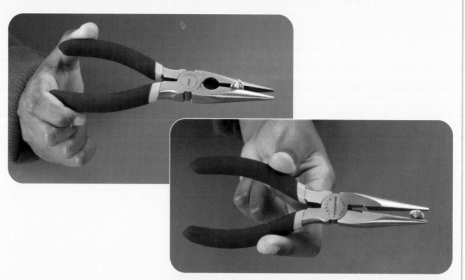

5 Try to spear a jellybean with the tongs and the pliers. Record your observations in your table.

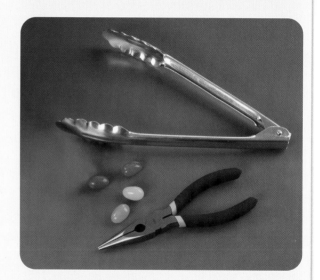

6 Use the tongs and the pliers to grab and pull out as many grass shoots or bean sprouts as possible. Record your observations in your table.

Analyze and Evaluate

1. Which tool is best for crushing a sunflower seed?

2. What was the best seed position for crushing it in the pliers—toward the tip or toward the handle? How does this relate to whether birds that eat seeds have a long beak or a short beak?

3. Birds that eat seeds need to have strong beaks. Which hand position represents a strong beak?

4. Examine the pictures of bird beaks in **Figure 1**. Which birds have beaks that are suitable for crushing seeds?

5. Which tool is best for spearing a jellybean?

6. Which tool worked best for grabbing and tearing grass shoots or bean sprouts? Which birds in **Figure 1** have beaks that are suitable for pulling grass and straining food from mud?

7. What conclusions can you make about the shape of a bird's beak and its feeding habits?

Apply and Extend

8. How is a beak that is suitable for spearing insects different from a beak that is suitable for spearing fish? Look at the bird beaks in **Figure 1**. Which birds have beaks suitable for spearing insects? Which birds have beaks suitable for spearing fish?

9. Look at the beaks of the pelican, the falcon, and the skimmer shown in **Figure 1**. What do you think each of these birds eat? Explain your thinking.

10. Look at **Figure 2**. This bird sucks the nectar from flowers. How does its beak help with this task?

Figure 2
A hummingbird

> ▶ *CHECK YOUR UNDERSTANDING*   ⊗
>
> **1.** What conclusions can you make about the shape of a beak and the size of the food that can be eaten?
>
> **2.** How did you use your table to help you organize and make sense of your data for this investigation?

4.3 Surviving in Extreme Conditions

▷ **LEARNING TIP**

Connect new information to what you have already learned. What structures and behaviours do you think would help organisms to survive in extreme conditions?

Some organisms have adaptations that enable them to survive Earth's most extreme conditions. For example, deep in the oceans, organisms can survive with little or no sunlight. Other organisms can live in dry deserts and in regions of extreme cold. What structures and behaviours enable them to survive such harsh conditions?

Canada's Arctic is home to many animals. In the winter, food is hard to find and temperatures may drop to −45 °C. Arctic animals have structures that allow them to survive in the cold. For example, the seal and the walrus have waterproof fur, the arctic grouse has fringed toes that act like a snowshoe, and the arctic fox has a thick white fur coat.

The polar bear has several structures that help it survive the Arctic cold. The polar bear has small, compact ears and a small tail, as well as thick fur. These adaptations help to keep it warm. The polar bear's white fur also helps camouflage the bear in the snow (**Figure 1**). This helps the polar bear sneak up on and hunt seals, as well as escape human hunters. The polar bear also has behaviours that help it survive in the winter. In the spring and summer, it eats as much as it can so that it has a thick layer of blubber when the winter comes. This blubber acts like insulation to protect the bear from the cold. (You will look at how insulation works in Unit C.)

Figure 1
A polar bear is well adapted to live in a snowy environment.

Skills Focus: observing, inferring

Look carefully at **Figures 2** and **3**. How have these animals adapted to winter?

Figure 2
A lynx

Figure 3
A snowshoe hare

Migration

Some animals have a behaviour that helps them survive the harsh winter. They move, or migrate, to a warmer place. This **migration** may not be a great distance. For example, the elk moves from the mountains to spend the winter in the lowlands. Other animals migrate great distances. For example, the humpback whale migrates from the Arctic region in the summer to the tropics in the winter. Other animals that take incredible migration journeys include the arctic tern and the Canada goose.

Long-Distance Travellers

The winner of the migration marathon is the arctic tern (**Figure 4**). This bird travels from the Canadian Arctic to Antarctica and back every year. Why does it make such a long journey? Does it need to fly this far for food and shelter? Most biologists believe that ancient relatives of the tern began making the journey when the continents were much closer together. Over millions of years, as the continents gradually shifted farther apart, the tern adapted to the ever-increasing distance of its migration.

Figure 4
The arctic tern migrates over 35 000 km each year.

The Canada goose is another long-distance traveller. It flies all the way to the southern United States and Mexico for the winter. Geese fly in a V-formation when migrating. Why do you think they fly in this formation? Think about the way you might shape your body if you wanted to travel fast. You would try to be streamlined. The lead goose hits the air with the greatest force. It breaks up the wind so that the wind flows with less resistance over the rest of the flock. Since the lead position is very tiring, the geese take turns being in the lead!

Hibernation

Other animals cope with winter by becoming inactive. This behaviour is called **hibernation.** Animals hibernate in burrows in the ground, in tree trunks, and in snow dens. Hibernating animals include chipmunks (**Figure 5**), some bats, and ground squirrels. When an animal hibernates, its body temperature drops and its heartbeat and breathing slow down. This allows the hibernating animal to use less energy so that it can live off the fat reserves it stored during the spring and summer. Some hibernating animals, such as chipmunks, also store food, such as nuts and seeds, to eat during the winter months.

Figure 5
The chipmunk spends the entire winter in its underground burrow. It wakes up now and then to eat part of the food it stored over summer.

Do you think of bears when you think of hibernation? In fact, bears are not true hibernators. Their body temperature does drop a few degrees, but they are easily awakened.

▶ CHECK YOUR UNDERSTANDING

1. Some people travel south for the winter. Are they migrating? Why or why not?
2. How does hibernating help an animal cope with the winter?
3. Draw an imaginary animal that would be well adapted to life in the Arctic. Explain your animal's adaptations.

Adapting to City Life  4.4

Organisms have adapted behaviours that help them to survive. Some animals have even learned how to survive in city environments (**Figure 1**).

Figure 1
What animals live in the city?

Today, humans are moving farther and farther into what used to be wild spaces. Houses, farms, roads, and shopping malls are replacing the natural habitats of animals. The animals have either moved to a new location or learned how to live with humans. Many animals have adapted to life surrounded by concrete, traffic noise, and a lot of people!

City Dwellers

A city is filled with roads with cars zooming by, tall buildings, and lots of people. This can be an advantage for some animals. Wherever there is traffic, buildings, and people, there is heat and food. The daytime heat becomes trapped between tall buildings and provides warmth for animals at night. The discarded food in a trash bin becomes a meal.

Figure 2
Birds, such as the house finch, have learned to build their nests in crevices of downtown buildings and apartment balconies.

The most successful city dwellers are birds (**Figure 2**). Pigeons are so common that in some cities they are considered to be a nuisance. Starlings have also learned to live in cities.

Many four-legged animals, such as squirrels and raccoons (**Figure 3**), have also learned how to live in cities. Raccoons can survive because they are willing to eat just about anything, from fresh vegetables in backyard gardens to waste in garbage cans. Rats and mice have also learned to live on the waste that is so easy to find in urban areas.

Small animals are not the only animals that live in cities. As cities expand, larger animals find themselves in direct contact with buildings and people. Seeing a cougar or a black bear in a backyard is not uncommon in parts of British Columbia.

Coyotes blend into the city so well that many people do not even know they are there (**Figure 4**)! Since coyotes are nocturnal, they roam the streets at night in search of food. Coyotes eat small mammals, such as rats, as well as eggs, fruit, grains, vegetation, and garbage. All these foods are easily found in a city!

Figure 3
Backyard compost bins, garbage cans, and even open back doors provide access to enough food for raccoons.

Figure 4
Coyotes are omnivores—they eat both plants and animals.

⫸ CHECK YOUR UNDERSTANDING ⊗

1. Name three animals that have adapted to live in cities. Describe how each animal has adapted.
2. What behaviours do pets, such as dogs and cats, learn to help them live in a home?
3. How can a city provide an advantage for an animal?

The Struggle Against Extinction

The diversity of life on Earth is amazing. Scientists believe that, of the total number of species that ever existed on Earth, most have died out, or have become extinct. Some extinctions are because of extreme environmental changes, such as ice ages and meteorite strikes. Groups of organisms disappear during these mass extinctions and are replaced with new species. This is what scientists believe happened to the dinosaurs 65 million years ago (**Figure 1**). Other species become extinct because they are not able to adapt to changing environments or changing food sources.

LEARNING TIP ◁

Check your understanding of the term extinction by describing in your own words what happens when an organism becomes extinct.

Figure 1
At the end of the Cretaceous Period, 65 million years ago, there was a mass extinction in which half of all life forms died out, including all the remaining dinosaurs.

Endangered Organisms

Many plants and animals are in danger of becoming extinct. These organisms are endangered. In Canada, 72 plant species and 95 animal species are endangered. Habitat destruction, hunting, and pollution are the main causes of species becoming endangered.

In Canada, we keep track of many organisms that are endangered. Environment Canada and the Canadian Wildlife Service are two agencies that help to manage and protect Canada's vast number of creatures. These agencies monitor organisms that seem to be disappearing. They conduct research into wildlife issues and work with other countries to preserve the world's diversity.

Let's look at three endangered species in British Columbia. As you read about these species, think about how their environment has been changed and what impact these changes have had.

The Tiger Salamander

The tiger salamander (**Figure 2**) lives in the often-dry Southern Okanagan Valley. Its habitat has been changed in two very important ways. First, fish that feed on the salamander have been introduced into some of the lakes where the salamander lives. Second, livestock that live in the area are trampling nearby plant life. This has affected the water quality of the lakes, making it more difficult for the salamander eggs to hatch. Both of these changes have affected the salamander's ability to survive.

Figure 2
The adult tiger salamander has large blotches of black, brown, and grey colouring along its body.

The Viceroy Butterfly

Figure 3
The Viceroy butterfly looks like the Monarch butterfly so birds do not eat it.

The Viceroy butterfly (**Figure 3**) lives in wetlands in southern Canada and throughout the United States. The Viceroy butterfly looks very much like the Monarch butterfly. **Mimicry** is an adaptation where an organism looks like another to help it survive. The Monarch butterfly tastes bitter and birds have learned not to eat it. Since the Viceroy looks like the Monarch butterfly, birds do not eat the Viceroy butterfly either.

The Viceroy butterfly feeds on the nectar of fruit trees. The butterfly is endangered in British Columbia because of the pesticides that fruit growers are using in their orchards. The pesticides kill the Viceroy butterfly along with the insects that harm the fruit.

The Vancouver Island Marmot

The Vancouver Island marmot (**Figure 4**) lives in the mountains of Vancouver Island. It likes alpine and sub-alpine areas, which have steep slopes, meadows, and rocky debris at the bases of cliffs. The Vancouver Island marmot is one of British Columbia's most endangered species. There are fewer than 50 left in the wild. The marmot is endangered because of loss of habitat. Although its habitat is now protected, the marmot is not reaching the suitable patches of land. Instead, it is remaining in areas that have been clearcut, where it is an easy target for predators, such as golden eagles, cougars, and wolves.

Figure 4
The Vancouver Island marmot is a small animal that lives in burrows. It eats grasses and other plants, and it hibernates during the winter.

TRY THIS: *IDENTIFY TRAITS*

Skills Focus: predicting, inferring

Look at the information about the three endangered species.

1. Identify a characteristic that has helped one of the species survive until now.
2. Think of another characteristic that would help this species survive in its changing environment.

▶ *CHECK YOUR UNDERSTANDING*

1. What is an endangered species?
2. Why do some organisms become endangered?
3. Choose one of the endangered animals discussed in the text. What do you think are some things that can be done to protect this animal?
4. Do you think that humans could become extinct? Explain your answer.

4 *Chapter Review*

Living things adapt to their environments.

Key Idea: Many organisms have structures that help them adapt to their environments.

Vocabulary

colouration
  p. 70

camouflage
  p. 70

mimicry p. 82

Key Idea: Many organisms have behaviours that help them adapt to their environments.

  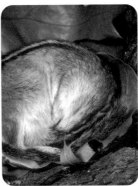

Vocabulary

behaviours p. 67

migration p. 77

hibernation
  p. 78

Key Idea: Environmental changes threaten some species with extinction.

Pesticides are endangering the Viceroy butterfly.

Review Key Ideas and Vocabulary

When answering the questions, remember to use the chapter vocabulary.

1. Organisms have adaptations that help them survive in their environments. Describe how each of the following adaptations helps the plant survive.
   - Dandelions have long taproots.
   - Arctic lupines have seeds that can wait for centuries before they sprout.
   - Lodge pole pine cones open only when there is enough heat.

2. Go back to **Table 1** in section 2.5 of Chapter 2, which lists the characteristics of vertebrates. For each class, name one characteristic that helps the vertebrates survive.

3. Which of the following are structures, and which are behaviours?
   - waterproof wings
   - eating berries
   - night vision
   - avoiding traps

4. Explain why the tiger salamander, the Viceroy butterfly, and the Vancouver Island marmot are endangered species. Are there any common reasons?

Use What You've Learned

5. Caribou migrate across the tundra every year. In some areas of the tundra, pipelines have been built to carry oil. These pipelines are barriers to the annual migration of the caribou. Explain how the caribou's behaviour must adapt to survive.

6. Earthworms live all across Canada in their underground environment. As they tunnel through the soil, they eat decaying organisms in the soil. They breathe directly through their thin skin. In dry areas, earthworms spend most of their time inside their burrows. If they didn't, their moist bodies would dry out and they would not be able to breathe. When it rains, earthworms come out of their burrows so they can mate. Explain how earthworms have adapted to their environment.

Think Critically

7. Sandpipers eat insects. Imagine that there is a nest of sandpiper chicks and that one of the chicks has a beak that is longer than the other chicks' beaks (**Figure 1**). How would this affect the chick's chances of survival?

Figure 1

8. List some human activities that may endanger the living things in your community. Which activity has the greatest impact? Explain your choice.

Reflect on Your Learning

9. Look back at the issues that were discussed in this chapter. Select the issue that is most interesting to you. Explain why you find this issue interesting. Describe something that you could do to have an effect on this issue.

Making Connections

Discovering Local Diversity

In this unit, you have learned about the different forms of life that exist on Earth. You have also learned how scientists classify these organisms, based on their similarities. You have used a microscope to examine organisms that seem to be invisible. As well, you have discovered how some organisms have evolved special characteristics and behaviours that help them survive.

Now it's time to apply what you have learned. In this investigation, you will create an inventory of the organisms that live in your community. Choose two different environments, perhaps a forested area and a grassy field.

Part 1: Conduct a Field Study

1. Carefully inspect each environment, and observe each organism you discover.

Remember to look closely and use your magnifying glass to see more. Make sure that you look everywhere—on the ground, in trees and shrubs, and above you in the sky. Gently lift rocks and scrape the surface of the soil to look underneath. Make sure that when you are finished, everything has been returned to the way you found it.

2. Record as much information as you can about the organisms you discover. Make descriptive notes about the environment that you find the organisms in and the number of each type of organism that you find.

3. Gather samples from each environment.
   - Collect water, if present.
   - Gather plant samples. Be careful to take your samples from debris on the ground. Do not break them off the plants.

Part 2: Identify Organisms

1. Classify the organisms as unicellular or multicellular. Then classify the organisms into the five kingdoms. Identify any members of the Animal kingdom as fish, amphibians, reptiles, birds, mammals, or insects.

2. Use a magnifying glass and microscope to investigate your samples. Record your observations.
   - Water samples: Prepare a slide to observe the organisms living in the water. Identify the kind and number of micro-organisms you see. What physical adaptations, such as a flagellum or cilia, do they have that help them survive? What behaviours, such as swimming fast or living in groups, do you observe that help them live in their environment?
   - Plant samples: Prepare slides of leaf and stem tissue for each plant. Look closely at the roots. What structures enable the organisms to survive in their environment? How are they able to meet their needs?

Part 3: Analyze Adaptations

1. Create a web for each environment that you investigated. In the centre, write a short description or draw a picture of the non-living parts of the environment (for example, a rocky, dry hill without any shade).

2. Around the centre, create a spoke for each organism that you discovered. Describe how the organism meets its needs in its environment. Give examples of any structures and behaviours the organism has that help it adapt to its environment.

3. Select one change (such as a drought) that could affect the organisms in each environment. Add another spoke to the web for this change, and describe adaptations that would help the organisms survive if this change occurred.

ASSESSMENT

Check to make sure that your work provides evidence that you are able to

CONDUCT A FIELD STUDY
- observe physical features of organisms and their environments
- use magnifying tools appropriately
- record accurate information about organisms and their environments in words and sketches
- conduct a field study without harming the environment

IDENTIFY ORGANISMS
- use magnifying tools appropriately
- classify organisms into five kingdoms
- distinguish among fish, amphibians, reptiles, birds, mammals
- use appropriate scientific terminology

ANALYZE ADAPTATIONS
- use magnifying tools appropriately
- identify structures and behaviours that are adaptations to the environment
- relate adaptations to specific features of an environment
- communicate clearly
- use appropriate scientific terminology

ELECTRICITY

Preview

Hey, what's going on? Without warning, the lights go out, your computer shuts down, your music stops, and your clock goes blank. No electricity! Quick, where are the candles?

Without electricity, your life would be different in just about every way. Imagine, for example, what your food would be like if you could not run all the appliances you use to prepare and cook it. Imagine not being able to listen to your music CDs or use the computer.

Look at this photo of North and South America. What does it tell you about the importance of electricity in our lives? Compare a brightly lit area with an unlit area. How would life be different in a place that has no lights? Find British Columbia on the map. Can you recognize some of the dots? Notice that not every place is brightly lit. Why are these areas not bright?

In this unit, you will discover what electricity is and the remarkable things it does for you. You will build electric circuits that can power light bulbs. You will also look at the production and transmission of electricity in British Columbia. As you learn about electricity, think about how much you depend on and use electricity, where you get the electricity you need, and why you need to conserve electricity.

TRY THIS: **USING ELECTRICITY**

Skills Focus: questioning, hypothesizing, inferring

Imagine how different your life would be if there were no electricity. Think about some of the electrical appliances in your home. Copy the following table into your notebook. In your table, list five appliances in your home. What purpose does each appliance have? What did people do before they had electricity to run this appliance? The first row is filled in as an example.

Appliance that uses electricity	Purpose of appliance	What people did before they had electricity
1. electric lamp	to provide light	used candles or went to bed when the Sun went down

◄ This is what North America and South America look like at night.

5

Electricity occurs when there is an imbalance of charged particles.

KEY IDEAS

▶ Everything contains charged particles.

▶ An electron is a negatively charged particle that forms part of an atom.

▶ An object is negatively charged when it has too many electrons and positively charged when it has too few electrons.

▶ Unlike electric charges attract, and like charges repel.

▶ Static electricity occurs when there is a buildup of electric charges on an object.

Have you ever seen a spectacular display of lightning, like the one in this photo? If you live in one of the drier parts of British Columbia, you've probably seen—and heard—a lot of electrical storms. Even if you have never seen lightning, it is occurring somewhere on Earth about 100 times every second!

When you think about electricity, you probably do not think about lightning. The electricity in lightning is an example of static electricity. What is static electricity? Is it the same as the electricity that runs appliances and makes life so convenient? In this chapter, you will discover what the lightning that flashes overhead has in common with the electricity that runs a CD player. Understanding electricity is one of the steps to understanding the world around you.

What Is Electricity?  5.1

TRY THIS: OBSERVE ELECTRICITY

Skills Focus: observing, communicating

Inflate a balloon, and put it against a wall. Watch what happens when you let the balloon go. Rub the same balloon with a wool cloth or your hair for a few seconds. Then put it against the wall. Observe what happens.

1. What happened to the balloon the first time?
2. What happened to the balloon the second time? How long did the balloon stay on the wall?
3. How can you explain what you observed?

When you think about electricity, you might think about plugging a hairdryer into a wall socket or running a CD player on batteries. However, you have experienced electricity in other ways. When your socks stick to each other in a clothes dryer, or when you rub a balloon on a wool sweater and then stick it to a wall, you are also experiencing a form of electricity (**Figure 1**).

To understand what electricity is, you have to look at **atoms.** Atoms are the tiny building blocks that make up everything around you—from the air you breathe to the clothes you wear. Everything is made of atoms. Each atom contains small particles that have an electric charge. Some particles have a **negative** charge (−). These particles are called **electrons** [ih-LEHK-trons]. Other particles have a **positive** charge (+). They are called protons. Since everything is made of atoms, everything contains charged particles.

Positive and negative charges pull, or attract each other. Most objects—including you, the chair you sit on, and this book—have an equal number of positive charges and negative charges. When the electric charges in an object are equal, or balanced, the object is neutral [NOO-truhl]. There are not any extra charges to attract something else. This is why a book does not stick to your hand and you do not stick to your chair.

So, why did the balloon that you rubbed with the wool cloth or your hair stick to the wall? What happens to make an object attract another object? The answer has to do with changing the electric charges in an object.

Figure 1

Thales, an Ancient Greek philosopher, noticed that when he rubbed amber (a yellow resin) with fur, it attracted feathers, threads, and leaves. The word "electricity" comes from the Greek word "*elektron*," which means "amber."

How Do Electric Charges Work?

We can change the balance of electric charges in an object so that the charges are unequal, or unbalanced. When this happens, we say that the object is charged with electricity.

For example, when you hold a balloon against a wall, it does not stick. It has the same number of positive and negative charges, so the balloon is neutral, or uncharged (**Figure 2**). Since the balloon is not attracted to the wall, it falls to the floor.

Neutral object

Figure 2
A neutral object, such as this balloon, has an equal number of positive charges (+ signs) and negative charges (− signs).

When you rub a wool cloth and a balloon together, both objects become charged. This does not mean that the rubbing creates the electric charges. The rubbing just moves the charges from their normal places. The rubbing knocks some electrons off the wool cloth, causing the cloth to change from being neutral to being positively charged. The balloon picks up the electrons and changes from being neutral to being negatively charged. **Figure 3** shows a positively charged ball and a negatively charged balloon.

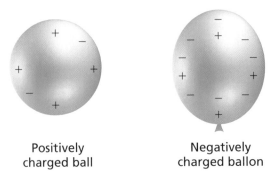

Positively
charged ball

Negatively
charged ballon

Figure 3
A positively charged ball has more positive charges than negative charges.
A negatively charged balloon has more negative charges than positive charges.

The Laws of Electric Charges

A positively charged wool cloth is attracted to a negatively charged balloon. Charged objects behave in certain ways according to the laws of electric charges. One of these laws states that objects with unlike charges attract one another (**Figure 5**). This means that if one object is negatively charged and another object is positively charged, they will move toward each other.

LEARNING TIP ◁

Counting the + signs and the − signs on the objects shown in the figures will help you see whether the object is positively charged, negatively charged, or neutral.

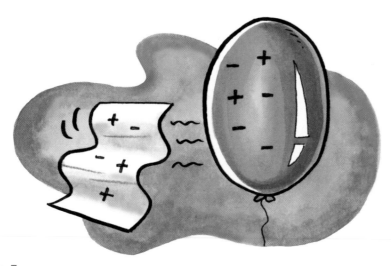

Figure 5
A negatively charged object attracts a positively charged object.

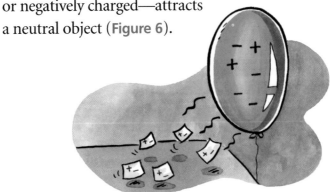

LEARNING TIP

The laws of electric charges are usually stated this way: Unlike charges attract, like charges repel.

Have you ever noticed that after you rub your hair with a balloon, the balloon can pick up small pieces of paper? Try it and see! A charged object—either positively charged or negatively charged—attracts a neutral object (**Figure 6**).

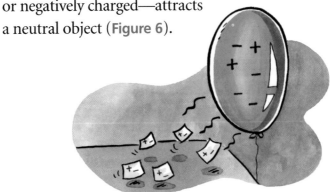

Figure 6
A negatively charged balloon attracts neutral pieces of paper.

Another law of electric charges states that objects with like charges repel each other. This means that two positively charged objects, or two negatively charged objects, will push away from each other (**Figure 7**). Can you use the laws of electric charges to explain what happened to the two balloons in the Try This activity?

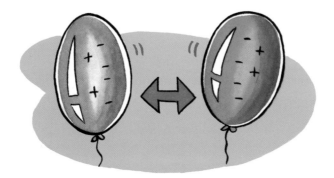

Figure 7
A negatively charged object repels another negatively charged object.

⫸ CHECK YOUR UNDERSTANDING ⊗

1. Use words and pictures to name and describe the two types of electric charges.

2. Read each description. Is the underlined object charged or neutral? Explain how you know.
   - Your <u>sock</u> is sticking to your shirt.
   - Two <u>books</u> are stacked on your desk.
   - <u>You</u> shuffle your feet across a carpet and get a zap when you touch a metal doorknob.
   - You are brushing your <u>hair</u> and notice that it begins to stand on end.

3. Use each pair of terms in a sentence.
   - attract, unlike charges
   - repel, like charges

What Exactly Is Electricity?

When you think about all the things you and your family do every day, it is easy to appreciate what electricity *does* to make your life easier. It is less easy, however, to understand what electricity is. To understand what electricity *is*, you have to understand the atom. Everything starts with the atom.

Atoms are tiny particles that make up everything in the universe. Each atom is made up of even tinier parts. In the centre, or nucleus, of an atom, there are small particles called neutrons and protons. Spinning around the nucleus are small particles called electrons (**Figure 1**). The electrons orbit the nucleus much like the Moon orbits Earth.

Figure 1
This atom has three protons, three electrons, and three neutrons. Protons have a positive charge. Electrons have a negative charge. Neutrons do not have a charge, so they are neutral. Each electron moves all around the nucleus.

An atom that has the same number of protons and electrons is neutral. Atoms tend to be neutral. When the number of protons and the number of electrons are different, the atom is charged. When an atom has a negative charge, the electrons move, or flow, to an atom with a positive charge (**Figure 2**).

Figure 2
Remember that opposites attract. A positively charged atom pulls, or attracts, a negatively charged electron.

How does this flow of electrons begin? One way to start a flow of electrons is to use friction. If you rub a balloon with a piece of wool, electrons from the wool rub off onto the balloon. Since the balloon now has more electrons, it is negatively charged and sticks to a neutral wall.

Another way to start a flow of electrons is to use a battery in a complete circuit. The electrons build up at the negative end of the battery and flow through a wire to a light bulb. The energy of the moving electrons powers the light bulb. The electrons then continue along another wire to the positive end of the battery (**Figure 3**). Thus, the electricity goes around and around the circuit.

Figure 3
Electrons move from the negative end of a battery to the positive end of the battery.

5.2 Static Electricity

There are two types of electricity: static electricity and current electricity. When the electric charges are in motion, the electricity is called current electricity. You will look at current electricity in Chapter 6. When the electric charges are at rest, or not moving, the electricity is called **static electricity.**

Static electricity occurs when electric charges build up on an object. This happens when negative charges are transferred from one object to another through rubbing, or friction. Examples of static electricity include the electric charges that build up when clothes rub against each other in a dryer, when you shuffle your feet across a carpet, or when you rub your hair or a pet's fur with a balloon (**Figure 1**). The effect of a wool cloth being rubbed on a balloon is another example of static electricity. Static electricity can even build up when gasoline flows through a nozzle or when an airplane travels through air. Static electricity is also responsible for the flashes of lightning that you see during a thunderstorm.

Figure 1
When a cat's fur is rubbed with a balloon, the balloon becomes negatively charged and the cat's fur becomes positively charged. The attraction between the balloon and the cat's fur results from static electricity.

Static electricity can be used to do work. For example, a photocopier uses static electricity to make copies. A Van de Graaf generator uses static electricity to produce some "hair-raising" effects (**Figure 2**).

Figure 2
Touching a Van de Graaf generator makes your hair stick out because each strand of hair has the same charge and is repelled by the other strands.

Where Do Static Electric Charges Go?

You probably noticed that a charged balloon does not stick to the wall for very long. What happens to the electric charges? Usually they go into the air, which is what happens to the charges on a balloon. When a balloon loses its static electric charges, it becomes neutral again and no longer sticks to the wall.

When static electricity leaves an object, it is called an electric discharge. An electric discharge can be slow and quiet, like when static electricity escapes into the air from a balloon. An electric discharge can also be sudden and can occur with a tiny spark, a shock, or a crackling sound, like when you pull a sock from a sweater. A lightning bolt is a dramatic example of an electric discharge (**Figure 3**). What words would you use to describe lightning?

Figure 3
Lightning can be very unpredictable. It is created when air currents, moving within large thunderclouds, push tiny ice crystals and water droplets past each other and transfer charges. When the static electricity is discharged, it produces both the flash of light and the sound of thunder.

▷ **LEARNING TIP**

Connect what you are reading to your daily life. Think about how you could prevent static electricity in your home.

Preventing Static Electricity

Everyone knows that lightning is dangerous. However, even small static charges can cause problems. For example, a static electric charge can damage the electronics of a computer. The charge is created as you touch your fingers to the keyboard and mouse. If you hold your hands close to a computer monitor, you may feel and hear the static electricity.

There are several ways to discharge charged objects. One way is to transfer the charges to the ground with a wire or a metal rod. This is called grounding. It is one of the simplest ways to discharge static electricity because Earth easily accepts electric charges.

People who work on sensitive electronic equipment wear a metal strip around one wrist. The metal strip is attached by a wire to a grounding system. This ensures that any charges are conducted, or transferred, to the ground.

You can discharge the static electricity in a computer, and avoid zapping the computer components, by placing the keyboard or computer on an anti-static mat, by touching the metal case of the computer or the metal legs of a chair, or by touching the floor and grounding yourself before working.

Tall buildings and some houses have grounding devices called lightning rods to protect them from lightning. A lightning rod is a metal rod that is attached to the highest part of a building. The rod is connected to a heavy copper wire that runs along the side of the building to a metal plate buried in the ground. If lightning strikes the building, the lightning rod directs the electric charge safely to the ground.

To prevent static electricity from building up in a clothes dryer, you can toss a fabric softener sheet in the dryer. The sheet will roll around with the clothes and lessen the buildup of electric charges.

CHECK YOUR UNDERSTANDING

1. Draw a diagram to show how static electricity can affect an object. Draw a second diagram to show one way that the static electricity on the object can be reduced.
2. Look at the following list and think about what you have learned:

   lightning bolt, grounding, discharge, zap, electric charge, attract, lightning rod, static electricity

   Find pairs of words that have some connection. Write the pairs in your notebook, and explain the connections. For example, "An electrician can use a metal strip around one wrist as a <u>grounding</u> device to help control <u>static electricity</u>."
3. Describe two ways that you can control a static discharge in your home.

5.3 *Design Your Own Experiment*

○ SKILLS MENU
○ Questioning   ● Observing
○ Predicting   ○ Measuring
● Hypothesizing   ○ Classifying
● Designing   ● Inferring
   Experiments
● Controlling   ● Interpreting
   Variables   Data
○ Creating   ● Communicating
   Models

Charging Materials

As you have learned, when you rub two neutral objects together, one object may become negatively charged and the other object may become positively charged. What kinds of materials will do this?

Design an experiment to test whether different pairs of materials become charged when rubbed together. For example, you may choose to test whether glass becomes charged when rubbed with a silk cloth, whether a straw becomes charged when rubbed through your hair, or whether a Styrofoam cup becomes charged when rubbed with a piece of wool cloth.

Question

Which pairs of materials will become charged when rubbed together?

Hypothesis

Write a hypothesis that states which pairs of materials will become charged when rubbed together. Make sure that you complete your hypothesis with a short explanation of your reasons. Write your hypothesis in the form "I think that … will become charged when rubbed together because …."

Materials

Decide which materials you will test. You can use **Figure 1** to get ideas. Be sure to test objects that you know are made from materials such as steel, iron, plastic, copper, silk, and wool.

> ▷ **LEARNING TIP**
>
> For help with writing a hypothesis or writing up your experiment, see "Hypothesizing" and "Writing A Lab Report" in the Skills Handbook. You may also want to review the "Designing Your Own Experiment" section.

Figure 1
What happens if you rub some of these materials together?

- Design a procedure to test your hypothesis. A procedure is a step-by-step description of how you will conduct your experiment. It must be clear enough for someone else to follow and do the exact same experiment.

- Submit your procedure, including any safety precautions, to your teacher for approval. Also, submit a diagram at least half a page in size, showing how you will set up your experiment.

Data and Observations

Design an observation table to record your observations.

Analysis

1. Which pairs of materials became charged when they were rubbed together? How do you know?

2. Which pairs of materials did not become charged when they were rubbed together? How do you know?

3. What is the same about the materials that became charged?

Conclusion

Look back at your hypothesis. Did your observations support, partly support, or not support your hypothesis? Write a conclusion for your experiment.

Applications

1. Would your results be different if the time between rubbing the objects and testing for a charge was longer or shorter? Design an experiment using one pair of materials that became charged, and include time as an independent variable.

CHECK YOUR UNDERSTANDING

1. Would you still have a fair test if you did not rub the pairs of materials the same number of times? Explain your answer.

2. Identify your independent, dependent, and controlled variables for this experiment.

Chapter Review

Electricity occurs when there is an imbalance of charged particles.

Key Idea: Everything contains charged particles.

A neutral object has an equal number of positive charges and negative charges.

Key Idea: An electron is a negatively charged particle that forms part of an atom.

Proton
Neutron
Electron
Nucleus

Vocabulary

atoms p. 91
electrons p. 91

Key Idea: An object is negatively charged when it has too many electrons and positively charged when it has too few electrons.

 Negatively charged ballon

 Positively charged ball

Vocabulary

negative p. 91
positive p. 91

Key Idea: Unlike electric charges attract, and like charges repel.

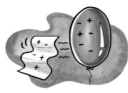

Key Idea: Static electricity occurs when there is a buildup of electric charges on an object.

Vocabulary

static electricity p. 96

Review Key Ideas and Vocabulary

When answering the questions, remember to use the chapter vocabulary.

1. What kinds of charged particles does an atom contain?

2. What is a negatively charged particle called?

3. Think about the number of negative and positive charges on an object. What can you say about an object that has a negative charge? What can you say about an object that has a positive charge?

4. Indicate whether each of the following objects would attract or repel each other:
   - an object with a positive charge and another object with a positive charge
   - an object with a positive charge and an object with a negative charge
   - an object with a negative charge and a neutral object

5. Suppose that you comb your hair using a plastic comb. When you finish, your hair is sticking to the comb. What has happened? How do you know?

6. List three ways that static electricity can be discharged. Give an example for each.

7. Explain where the lightning in the picture is likely to strike.

Use What You've Learned

8. If you live in a dry area of British Columbia, you may see small sparks when you move your feet back and forth across your sheets at night. Explain what is happening.

9. Would it be better to wrap a sandwich in plastic wrap or wax paper? Explain why, in terms of static electricity.

10. Explain why static electricity is an uncontrolled form of electricity.

Think Critically

11. How could you avoid getting a shock from static electricity when you touch a metal doorknob?

12. List three safety precautions that you should take if you are outside during a lightning storm.

Reflect on Your Learning

13. What questions do you still have about static electricity? Do you think your questions will be answered in the rest of this unit? If not, where can you go to find the answers?

CHAPTER 6

Current electricity can be controlled and directed.

▐▶ KEY IDEAS

▸ An electric current is the flow of electrons.

▸ A complete circuit has three parts: a source of electricity, a pathway, and an electrical device.

▸ Conductors permit the flow of electric current, and insulators resist the flow of electric current.

▸ Electricity may flow in series circuits or parallel circuits.

As you have learned, you can produce static electricity by rubbing two materials together. The static electric charge builds up until it discharges in the form of a small spark. However, computers, televisions, and electric lights, like the ones shown in the above photo, do not run on static electricity. They need a continuous flow of electricity.

You can think of electricity flowing along a pathway as being like water flowing through a hose. Just like water, you can direct the electricity to where you want it to go, and you can turn it on and off whenever you want. Being able to direct and control electricity makes it very useful to us.

How Electricity Flows along a Circuit

Every electrical device that you use—from a hair dryer to a computer—requires a continuous flow of electrons moving from one place to another along a pathway. The term **current electricity** refers to electricity that is produced by the flow of electrons along a pathway. The flow of electrons is called an **electric current.**

For an electric current to flow, you need a **circuit** [SUR-kit]. A complete circuit has a source of electricity, such as a battery or a generator, a pathway for the electrons to flow through, and an electrical device to operate, such as a light bulb (**Figure 1**).

> **LEARNING TIP** ◁
>
> Compare what you are learning with what you already know. As you read this chapter, ask yourself how current electricity fits with what you have learned about static electricity. How are they the same and how are they different?

Figure 1
The continuous flow of electricity around a circuit lights up these strings of decorative lights.

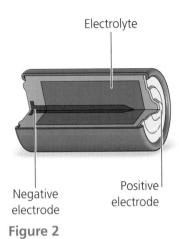

Electrolyte

Negative electrode

Positive electrode

Figure 2
The inside of a cell

Source of Electricity

The electricity source can be a **battery** or a generator. A battery uses a chemical reaction to create an electric current. A battery produces electric current that flows in one direction only. This current is called **direct current.**

A battery is really two or more electric cells connected together. For example, some electronic games use a battery of two cells, while flashlights use a battery of four cells. However, we commonly use the term battery to refer to one cell. All electric cells contain two metal plates, called electrodes, in a chemical mixture called an electrolyte [ih-LEHK-truh-lite] (**Figure 2**). The mixture causes one electrode to produce electrons and the other electrode to accept the electrons. When the electrodes are connected to a device, such as a light bulb, the electrons flow from the negative end of the battery around the circuit to the positive end.

Have you noticed that the family car has a 12-V (volt) battery, while an electronic game needs 1.5-V batteries? The numbers 12 and 1.5 represent the **voltage** of each battery. They refer to the push the batteries give to the electrons to move the electrons along a circuit.

TRY THIS: *MAKE A LEMON CELL*

Skills Focus: questioning, observing, interpreting data, communicating

 Wear safety goggles during this activity.

You can make your own cell using a lemon, a 10-cm piece of copper wire, a straightened paper clip, and a digital multimeter set to measure current in millivolts (mV). First, roll the lemon around on a hard surface to loosen the juices inside. Then stick about 3 to 4 cm of copper wire into the lemon. Put 3 to 4 cm of the paper clip into the lemon about 1 cm from the copper wire. Touch one probe of the meter to the copper wire and the other probe to the paper clip (**Figure 3**). What happens? Reverse the position of the probes. What happens?

1. What do you observe? Explain.
2. Replace the paper clip with another copper wire. What happens?
3. Replace the copper wires with two paper clips. What happens?

Figure 3

Circuit Pathway

A circuit needs not only a source of electricity, such as a battery, but also a circuit pathway that the current can flow through. A circuit pathway is usually a wire.

Materials that let electricity flow through them easily are called **conductors.** All metals are good conductors. For that reason, a circuit pathway is usually a copper or steel wire.

Other materials, such as plastic and wood, do not allow electricity to flow easily. These materials are called **insulators.** Insulators resist the flow of electric current.

Electrical Device

A circuit, with a source of electricity (a battery) and a pathway for the current, is used to run an electrical device. An electrical device changes electricity into light, or heat, or the form of energy that is needed. Appliances such as stoves and refrigerators, computers, televisions, and light bulbs (**Figure 4**) are all electrical devices.

LEARNING TIP ◁ ▶

As you read, ask yourself questions to check your understanding: What did I just read? What did it mean? For example, check your understanding of the highlighted words on this page by defining them in your own words.

Figure 4
A light bulb is an electrical device. It changes electricity into light and heat.

Switches

Once you have a circuit, you can control the flow of electric current using a device called a **switch.** When the switch is on, or closed, the circuit is complete and the current can flow along the pathway. This type of circuit is called a **closed circuit.** When the switch is off, or open, the circuit is incomplete and the current cannot flow. This is called an **open circuit** (Figure 5).

▷ **LEARNING TIP**

As you study **Figure 5**, ask yourself: What is the purpose of this diagram? Why is it included? What am I supposed to notice or remember?

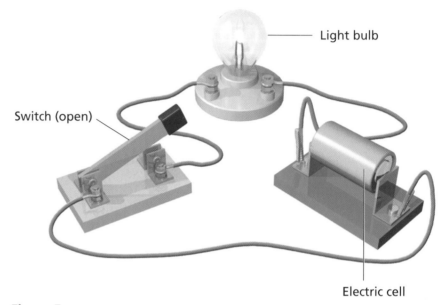

Light bulb

Switch (open)

Electric cell

Figure 5
Current cannot flow in this open circuit.

TRY THIS: *MAKE A CIRCUIT*

Skills Focus: observing, communicating

Use a light bulb in a bulb holder, two connecting wires, and a cell in a holder to make your own electric circuit. Draw a diagram of your circuit. Label each part of your circuit and explain its purpose.

LEARNING TIP ◁

Do not guess when answering questions. Look back through the section to find the answers. Even if you remember the answer, it is good to go back and check.

1. What are the three parts of a complete electric circuit? Draw and label a diagram.

2. Explain why a television repairperson and an auto mechanic need to know how to determine whether or not a circuit is complete.

3. Think about a battery-powered device. Why do you need to put the batteries in the correct positions?

4. Look at the four circuits shown in **Figure 6**. Which of the circuits is (are) complete?

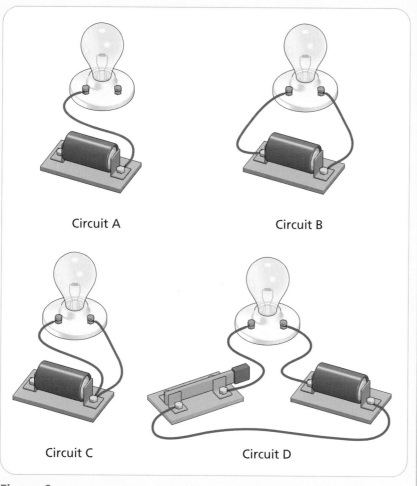

Circuit A

Circuit B

Circuit C

Circuit D

Figure 6
Circuits for Question 4

6.2 Series Circuits

When a circuit has only one pathway for the current to travel through, as shown in **Figure 1**, it is called a series circuit. A series circuit is like a track. All the electrons have to follow the same pathway, just like all the runners have to follow the same path. The pathway forms a complete loop from the source, through one of the wires to an electrical device and then through another wire back to the power source. Some electrical devices, such as light bulbs, can be connected to one another. The electric current passes through all the electrical devices on the series circuit when the circuit is closed.

Figure 1
A series circuit has only one pathway for the current to flow through.

TRY THIS: **MAKE A SWITCH**

Skills Focus: predicting

How can you design a switch using the materials shown in **Figure 2**? Make a simple switch to add to a series circuit. When you close the switch, the electric current should flow along the pathway to light the bulb.

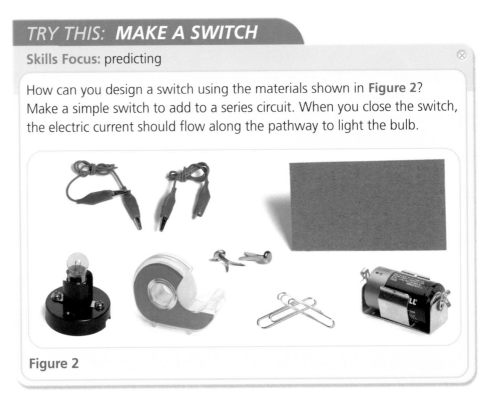

Figure 2

Drawing a Circuit

How do electricians know how to put together circuits so that electrical devices will work? They first draw plans, called schematic circuit diagrams, of the circuits. A schematic circuit diagram has symbols, instead of words or pictures, for different parts of the circuit so that the circuit is easy to draw and understand. **Table 1** shows some circuit diagram symbols.

LEARNING TIP ◁

We often use the term battery instead of cell. A battery is actually a combination of two or more cells. The circuit diagram symbol for battery shows three cells.

Table 1   Circuit Diagram Symbols

Part of circuit	Symbol
cell	—+‖—
battery (3 cells)	—+‖‖‖—
conducting wire	——
light bulb	—⊗—
open switch	—∘⁄∘—
closed switch	—∘∘—

Figure 3 shows a schematic circuit diagram of a series circuit that has three light bulbs connected to a cell. The arrows indicate the pathway that the electrons take through the light bulbs.

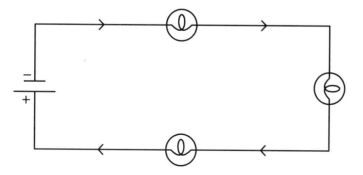

Figure 3
A schematic diagram of a series circuit

LEARNING TIP ◁

To read the diagram in **Figure 3**, first look at the symbols in **Table 1**. Then follow the path of the arrows in the diagram. Check that you can see where each part of the circuit is on the diagram.

▶ CHECK YOUR UNDERSTANDING ⊗

1. Create a schematic circuit diagram for the circuit shown in **Figure 1**.

2. How does a series circuit work?

▶ SKILLS MENU

○ Questioning	● Observing
● Predicting	○ Measuring
○ Hypothesizing	○ Classifying
○ Designing Experiments	● Inferring
○ Controlling Variables	○ Interpreting Data
○ Creating Models	● Communicating

Materials That Conduct Electricity

As you have learned, a circuit requires a conducting pathway. Generally, metal wires are used to connect a device to an electricity source and to complete the circuit. Why are the wires usually coated with plastic?

Electricity can flow easily through conductors, such as metals. Insulators, such as plastic and wood, resist the flow of electric current. In this investigation, you will identify materials as either conductors or insulators.

Question

Which materials conduct electricity?

Materials

- cell (battery) in a holder
- 3 connecting wires
- light bulb in a bulb holder
- different materials to test, such as a metal fork, a pencil, a copper wire, a piece of aluminum foil, an eraser, a paper clip, a plastic straw, a potato, an orange, and a thick rubber band

cell (battery)

wires

light bulb

metal fork

paper clips

pencil

eraser

copper wire

thick rubber band

aluminum foil

potato

plastic straw

orange

▶ Procedure

1 Copy the following table into your notebook. List all the materials you will be testing. Predict whether each material is a conductor or an insulator.

Observations for Investigation 6.3		
Material	Prediction	Observations
metal fork		

2 Build a series circuit. Remember to connect the wires to the battery last.

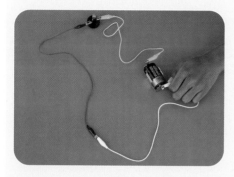

🖐 Do not touch the two ends of the wire together when they are connected to the battery.

3 Disconnect the wires from the battery. Connect the wires to the material to be tested, and then reconnect the wires to the battery. Record your observations in your table.

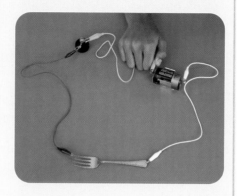

4 Repeat step 3 until you have tested all the materials.

Analyze and Evaluate

1. Which materials were the best conductors? What characteristics do these materials share?

2. Which materials were the best insulators? What characteristics do they share?

3. Is there any variable that could have affected your results? If so, how would you change the investigation to control this variable?

Apply and Extend

4. When would it be important for someone to understand the difference between an insulator and a conductor?

5. Is a resistor a conductor or an insulator? Explain your thinking.

�III▶ CHECK YOUR UNDERSTANDING ⊗

1. In step 3 of the procedure, you were asked to record your observations in your table. Why is it important to include your observations, rather than just recording whether or not the light bulb came on?

2. Why did you disconnect the wires to the battery before taking the two wires apart and connecting them to the material you were testing?

A circuit that has more than one pathway for the current to flow through is called a **parallel circuit** (**Figure 1**). You can think of a parallel circuit as being like the streets in your neighbourhood. Just as you can travel along different streets to get home, the current can travel along different pathways to return to the source.

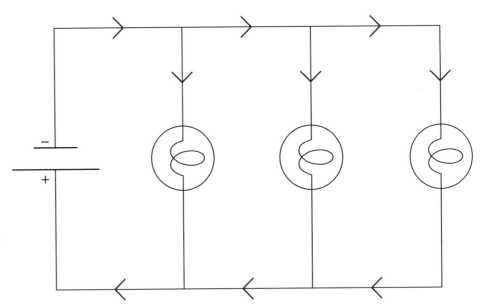

Figure 1
A parallel circuit has different pathways that the electric current can take. The total current will divide and flow through each of the three pathways, lighting each light bulb before recombining and returning to the source.

TRY THIS: BUILD A PARALLEL CIRCUIT

Skills Focus: observing, communicating

Look carefully at the parallel circuit shown in **Figure 1**.

1. Build the parallel circuit using connecting wires, a cell (battery) in a holder, and three light bulbs in bulb holders.
2. Describe how the current flows in this circuit.

> **LEARNING TIP**
>
> Check your understanding of series circuits and parallel circuits by explaining the difference to a classmate.

In a parallel circuit, the electric current divides and flows through all the pathways to the light bulbs. Since each light bulb in a parallel circuit is connected to its own separate pathway, what do you think would happen if one light bulb was removed? In the next investigation, you will find out.

114 Unit B **Electricity**

NEL

1. Think about what you have learned about a series circuit and a parallel circuit. How are the circuits the same? How are they different?

2. Which of the circuits shown in **Figure 2** is a parallel circuit?

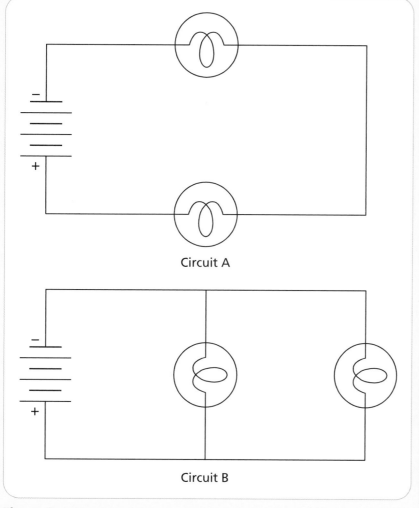

Circuit A

Circuit B

Figure 2
Circuits for Question 2

Conduct an Investigation

○ SKILLS MENU

○ Questioning	● Observing
● Predicting	○ Measuring
○ Hypothesizing	○ Classifying
○ Designing Experiments	● Inferring
○ Controlling Variables	○ Interpreting Data
○ Creating Models	● Communicating

Removing a Light Bulb

You have learned how electricity flows through a series circuit. You have also learned how electricity flows through a parallel circuit. Suppose that you want to light two light bulbs using one battery. In this investigation, you will compare a series circuit that contains two light bulbs with a parallel circuit that contains two light bulbs. As well, you will discover what happens when a light bulb is removed from each circuit.

Question

What happens when a light bulb is removed from a series circuit and a parallel circuit?

Predictions

i) Predict what will happen when one light bulb is removed from a series circuit.

ii) Predict what will happen when one light bulb is removed from a parallel circuit.

light bulbs

Materials

- 2 light bulbs, each in a bulb holder
- 4 connecting wires
- cell (battery) in a holder

wires

cell (battery)

▶ Procedure

1 Build a series circuit with two light bulbs as shown in the photo.

2 Test the circuit to make sure that it is complete.

🖐 Do not touch the two ends of the wire together when they are connected to the battery.

3 Disconnect the battery, and remove one light bulb from the circuit.

4 Reconnect the battery, and observe what happens. Record your observations.

5 Build a parallel circuit with two light bulbs as shown in the photo.

6 Repeat steps 2 to 4 for the parallel circuit.

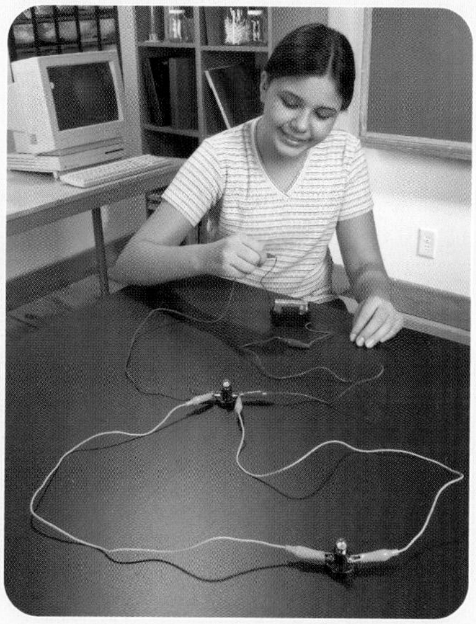

Analyze and Evaluate

1. What happened when you removed a light bulb from the series circuit? Explain why.
2. What happened when you removed a light bulb from the parallel circuit? Explain why.
3. Were your predictions correct?

Apply and Extend

4. What is an advantage of a parallel circuit?
5. Why are the electric circuits in your home parallel circuits?
6. Look at **Figure 1**. Which object do you think contains a parallel circuit? Explain your reasoning.

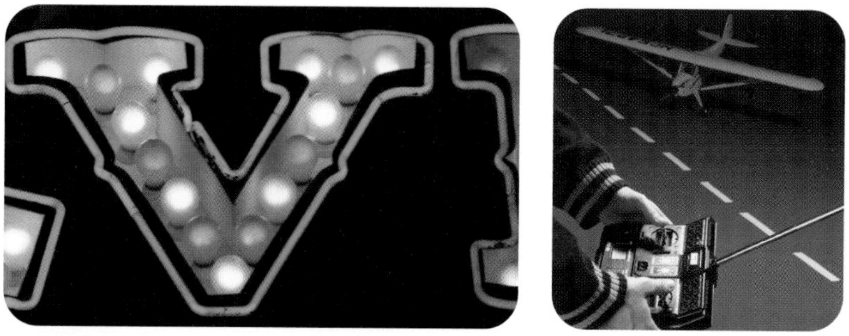

Figure 1
An electric sign and a battery-operated toy

CHECK YOUR UNDERSTANDING

1. An inference is an interpretation of available evidence. In this investigation, you used your observations to make inferences. You explained why removing a light bulb had certain effects. Are inferences always correct? Can you make an inference without first making observations or collecting data? How can you check to see if an inference you have made is correct?

6 Chapter Review

Current electricity can be controlled and directed.

Key Idea: An electric current is the flow of electrons.

Vocabulary

current electricity
  p. 105

electric current
  p. 105

Key Idea: A complete circuit has three parts: a source of electricity, a pathway, and an electrical device.

This circuit contains a cell (battery), wires, and a light bulb.

Vocabulary

circuit p. 105

battery p. 106

direct current p. 106

voltage p. 106

switch p. 108

closed circuit p. 108

open circuit p. 108

Key Idea: Conductors permit the flow of electric current, and insulators resist the flow of electric current.

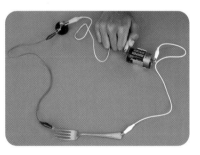

Metals are conductors.

Vocabulary

conductors p. 107

insulators p. 107

Key Idea: Electricity may flow in series circuits or parallel circuits.

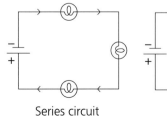

Series circuit

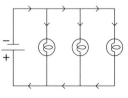

Parallel circuit

Vocabulary

series circuit p. 110

parallel circuit
  p. 114

Review Key Ideas and Vocabulary

When answering the questions, remember to use the chapter vocabulary.

1. Describe the difference between static electricity and current electricity.

2. What are three parts of an electric circuit?

3. What is the difference between a conductor and an insulator? Where do we use a conductor? An insulator?

4. Create two schematic circuit diagrams, one of a parallel circuit and one of a series circuit, each with two bulbs, a cell (battery), and an open switch.

Use What You've Learned

5. Explain how a switch controls the flow of electric current. Use a schematic circuit diagram in your explanation.

6. Why can static electricity not be used to run a television?

7. Suppose that you have just found an unknown material in a cupboard. How can you find out if the material is a good conductor or insulator of electricity?

8. Copper wire is a good conductor of electricity. Do you think that the thickness of the wire will affect the brightness of the light bulb? Design an experiment to find out.

9. You want to invent an electrical device that will transport you to school with a minimum of effort on your part. Where will you use a conductor in your design? Where will you use an insulator? Explain why. Will you use series circuits or parallel circuits, or both? Explain why. Sketch your design.

10. Most bedrooms have two or more outlets for plugging in electrical devices. Would a bedroom be wired using a series circuit or a parallel circuit? Explain why.

11. Imagine that you are an electrician. You have been called to find out why the lights in a kitchen will not work. What are you going to look for to solve the problem?

Think Critically

12. Why are electrical wires coated in rubber or plastic?

Reflect on Your Learning

13. You have probably flicked at least one switch today to turn on a light or an appliance. How has learning about different kinds of circuits made you appreciate how electricity works?

CHAPTER 7

Electricity can be changed into other forms of energy.

▶ **KEY IDEAS**

▸ Electricity can be transformed into other forms of energy, such as light, heat, motion, and sound energy.

▸ Electric currents produce magnetic fields.

▷ **LEARNING TIP**

Before reading this chapter, take a quick look through it, noticing the headings, the illustrations, and the activities. Make a list of predictions about what you will learn in this chapter.

When you turn on a lamp, you are seeing light made with electricity. When you cook your lunch on an electric stove, you are using heat made with electricity. When you play a CD or listen to a radio, you are hearing sounds that were produced using electricity. A motor, like the one from the disk drive of a computer shown above, uses electricity to produce motion. Electricity can be changed into different forms of energy that we can use.

In this chapter, you will look at how electricity is changed into different forms of energy. You will also discover how electricity is used to produce a magnetic force.

Transforming Electricity into Light and Heat

Electricity is most commonly used to produce light. With a flick of a light switch, electricity is sent to a **light bulb** and transformed, or changed, into light and heat. An incandescent [IN-kuhn-DES-uhnt] light bulb produces light by heating a filament [FILL-uh-muhnt] (**Figure 1**). Did you know that electric lighting has only been around since the early 1880s?

Tungsten wire filament

Glass bulb

Figure 1
The incandescent light bulb is used in lamps to produce light. When electricity passes through the tungsten wire filament, it becomes so hot that it glows white.

How is electricity transformed into light in a light bulb? The filament in a light bulb resists the flow of electric current. Like a crowd of people trying to get through a narrow hall, the electric charges meet resistance as they travel along the filament. This makes the filament so hot that it turns red and then white. The extreme white heat creates the light that you see. The filament does not burn because it is made of a tough metal called tungsten. The glass bulb surrounding the filament is filled with a gas that does not allow burning. Eventually, the filament gets so thin that it breaks and the bulb has to be replaced.

Have you ever touched a light bulb when it was on? If you have, you know that not all the electricity is changed into light. Some of the electricity is changed into heat. This heat is wasted. It goes into the air around the light bulb and is not used for anything (**Figure 2**).

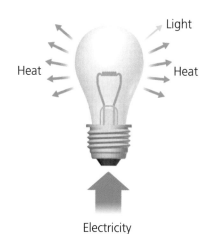

Light

Heat

Heat

Electricity

Figure 2
Only about 6% of the electricity that goes into a light bulb produces light. The rest produces heat.

Fluorescent [flo-REHS-uhnt] light bulbs also use electricity to produce light. A fluorescent light bulb is a glass tube that contains gases. These gases produce a special type of light, called ultraviolet light, when electricity is passed through them. A coating inside the fluorescent tube absorbs the ultraviolet light and produces the light that you see. Fluorescent light bulbs use a lot less electricity and last much longer than incandescent light bulbs.

Light emitting diodes, or LEDs, are used in digital clocks, bicycle lights, Christmas lights, newer traffic lights, and calculators. LEDs are tiny solid light bulbs that are very energy-efficient (**Figure 3**). They produce very little heat because they do not have a filament. As well, they last much longer than other types of light bulbs. They are expensive, however. As the price of LEDs falls, more and more electrical appliances will use them.

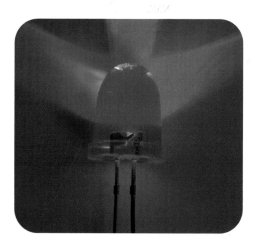

Figure 3
An LED bulb is more durable than an incandescent light bulb because it has no filament and no glass bulb that can break.

Some appliances, such as toasters, heaters, and hair dryers, are designed to produce heat. The wires inside a toaster are similar to the filaments in a light bulb. They resist the electric current and produce heat. This heat toasts your bread. The wires also produce light—the orange or red glow.

Ⅲ▶ CHECK YOUR UNDERSTANDING ⊗

1. When electricity is transformed into light, another form of energy is produced. What is it?

2. Why do many people now use fluorescent light bulbs instead of incandescent light bulbs?

3. What are three advantages of LEDs?

Using Electricity to Produce Magnetism

Magnets exert a force. The area of force around a magnet is called the magnetic field. The magnetic field is strongest at the poles of the magnet. As you saw in the activity, unlike poles attract each other and like poles repel each other. In this way, magnets are just like electric charges.

   About 2500 years ago, people noticed that some rocks, called lodestones, attracted small pieces of iron (**Figure 1**). We now call the property of attracting or repelling iron **magnetism.** Scientists have found that electric currents produce magnetic fields the same way that magnets do. In other words, electricity produces magnetism. We call the magnetic forces produced by electricity **electromagnetism.**

LEARNING TIP ◁

Magnetite was first discovered near Magnesia, Greece. This is where the word "magnetism" comes from.

Figure 1
A piece of magnetite that is magnetic is called a lodestone, or "leading stone," because it points north and can be used to lead the way home.

Skills Focus: creating models, observing

Use sandpaper to remove the insulation from the ends of an insulated wire. Wrap the wire in tight, even coils, as many times as you can around a long iron nail. Leave about 10 cm of wire at both ends of the nail to attach to a battery. Attach the ends of the wire to the battery. Hold the nail above a pile of paper clips. Observe what happens. Disconnect one of the wires from the battery. Hold the nail above the paper clips again. Observe what happens.

1. What does this activity show you about electricity and magnetism?

 Do not leave the electromagnet connected for too long. It will become very hot.

Magnets that are created using electricity are called electromagnets (**Figure 2**). An electromagnet has a source of electricity, such as a battery, and a wire coiled around an iron object, called a core. An electromagnet works as long as the circuit is closed and the electric current is flowing. When the circuit is open, there is no attraction. This means that an electromagnet can be turned on and off.

Figure 2
Huge electromagnets are attached to cranes and used to separate iron and steel scrap in recycling plants. These electromagnets are even strong enough to pick up an entire car!

Did you know that there are many electromagnets in your home? Doorbells, telephones, and electric motors all contain electromagnets.

Electromagnets are also being used to develop high-speed trains, called magnetic levitation, or maglev, trains (**Figure 3**). These trains do not run on tracks, but use electromagnetism to float above the track.

Figure 3
A maglev train travels at speeds up to 500 km/h along a test track in Japan.

Electricity and magnetism are related. Electricity moving through a wire produces a magnet. Similarly, a magnet moving inside a coil of wire produces an electric current in the wire. Electrical generators use the principle of electromagnetism to produce electricity. In Chapter 8, you will learn more about how electricity is generated.

⫸ CHECK YOUR UNDERSTANDING

1. How can you use electric currents to produce magnetism?
2. What is the difference between a permanent magnet and an electromagnet? What is the advantage of using an electromagnet over using a permanent magnet?
3. How would your day be different without electromagnetism? Draw two pictures that show what your life would be like with and without electromagnetism.

SKILLS MENU

- ○ Questioning
- ● Observing
- ● Predicting
- ● Measuring
- ○ Hypothesizing
- ○ Classifying
- ○ Designing Experiments
- ● Inferring
- ○ Controlling Variables
- ○ Interpreting Data
- ○ Creating Models
- ● Communicating

Making a Stronger Electromagnet

You have learned that electromagnets can be created using electricity. You have also learned that an electromagnet consists of a source of electricity and wire coiled around an iron core. In section 7.2, you made a simple electromagnet using a battery, a coiled wire, and an iron nail. How can you build a stronger electromagnet? Does the number of coils affect the strength of an electromagnet?

Question

How does the number of coils affect the strength of an electromagnet?

Prediction

Use what you have learned about magnetism to predict how the number of coils will affect the strength of an electromagnet.

Materials

- sandpaper
- insulated wire, about 1 m long
- cell (battery) in a holder
- uncoated paper clips
- long iron nail or spike
- ruler

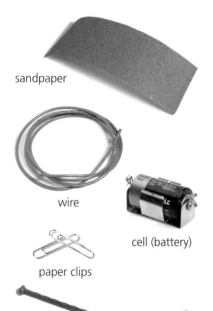

sandpaper

wire

cell (battery)

paper clips

nail

ruler

▶ Procedure

1 Copy the table below into your notebook. Record your observations in this table.

Observations for Investigation 7.3

Number of coils	Number of paper clips picked up
10	
20	
30	
40	
50	

2 Use sandpaper to remove the insulation from the ends of the wire. Wrap the wire around the nail core 10 times in the same direction. Make sure that there is enough wire at each end to connect to the battery.

 Do not leave the electromagnet connected for very long. It will become very hot.

3 Connect one end of the wire to the battery. Hold the end of the nail 1 to 2 cm above a pile of paper clips.

4 Connect the other end of the wire to the battery. Record the number of paper clips that are attracted to your electromagnet.

5 Disconnect the battery. Wrap 10 more coils of wire around the nail. Connect one end of the wire to the battery. Hold the end of the

nail 1 to 2 cm above a pile of uncoated paper clips. Connect the battery. Record the number of paper clips that are attracted to your electromagnet.

6 Repeat step 5 until you have 50 coils of wire around the nail.

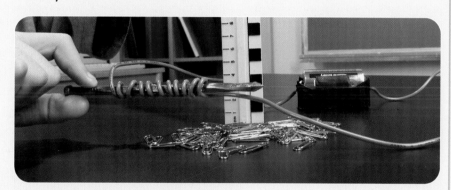

Analyze and Evaluate

1. Create a bar graph using the results in your observation table.

2. What happened to the strength of the electromagnet as the number of coils increased?

3. How well did you predict the results of this investigation?

LEARNING TIP ◁

To review bar graphs, see the Skills Handbook section "Graphing Data."

Apply and Extend

4. Predict the strength of an electromagnet that has 100 coils. Add your prediction to your graph.

5. What other variable could affect the strength of the electromagnet? How would you design an investigation to determine the effect of this variable?

⫸ CHECK YOUR UNDERSTANDING

1. What was the independent variable in this investigation? What was the dependent variable?

2. List at least three variables that you had to control to ensure that your test was fair.

Transforming Electricity into Motion

Skills Focus: observing

An electric motor converts electricity into motion. Can you use a small motor to create an optical illusion that will mesmerize your partner?

Obtain a motor from your teacher. Cut a circle out of thin cardboard and attach it to the shaft of the motor. Draw a pinwheel pattern on a white piece of paper and tape it to be cardboard base. Connect the motor to a battery and a switch using connecting wires (**Figure 1**). Observe what happens when you close the circuit.

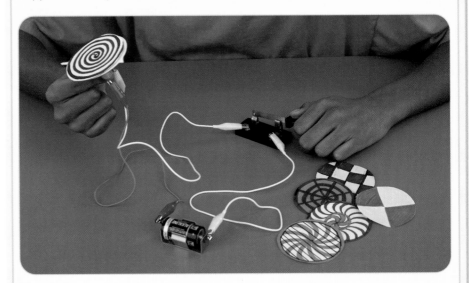

Figure 1

1. Have your partner look at the pinwheel pattern while the circuit is closed. What does your partner see?
2. Switch places with your partner. Describe what you see when you look at the pinwheel pattern when the circuit is closed.

▷ **LEARNING TIP**

The words "motor" and "motion" both come from the Latin word meaning "move."

If you looked around your home, you would find many electric motors. You would find them in small appliances, such as a blender, a hair dryer, and the disk drives in a computer or CD player. You would also find motors in large appliances, such as a refrigerator (**Figure 2**), a dishwasher, and a washing machine. There are also electric motors in most electric tools, such as drills, saws, sanders, and planers. The windshield wipers on your family car are powered by an electric motor, as is an entire electric trolley bus (**Figure 3**). If something electric has a part that moves, it probably has a motor.

Figure 2
The electric motor of your refrigerator powers a compressor to keep your food cold.

Figure 3
An electric trolley bus gets its power from overhead electrical cables.

A motor transforms electricity into motion. Although a motor uses electricity, it actually works using electromagnetism. Inside a motor is a permanent magnet and an electromagnet, which can rotate. When the circuit is closed, the electromagnet is constantly repelled and attracted to the permanent magnet. This causes the electromagnet to turn. This also causes anything that is attached to the electromagnet to turn, causing motion. When the circuit is open, the motor stops working because the coil is no longer an electromagnet.

▸ CHECK YOUR UNDERSTANDING

1. How is electricity transformed into motion?
2. What appliances in your home transform electricity into motion?

7.5 Transforming Electricity into Sound

Sound is created when something vibrates, or moves back and forth quickly. Think about a guitar string. If you pluck a guitar string, it vibrates and causes the air around it to vibrate. The vibrating air, or sound wave, travels to your ear and makes your eardrum vibrate. This is how you hear the sound!

Electricity can be transformed into sound (**Figure 1**). When lightning strikes, for example, a massive electric charge shoots through and heats the air around it. This intense heat causes the air to expand and then contract, creating an explosion of sound that you hear as thunder.

Figure 1

When you listen to music on a CD player, you are listening to sounds made with electricity. Sounds are first changed into a code number. A laser then burns pits into the reflective layer of a CD according to the code. A laser in your CD player reads the code, and the code is turned back into the different sounds you hear.

Electricity can also be transformed into sound using electromagnets. **Figure 2** shows how a doorbell works. When the button is pressed, the circuit is closed. Current flows through the coils, creating an electromagnet. The electromagnet attracts the arm, which is attached to a hammer. This makes the hammer strike the bell, producing a sound.

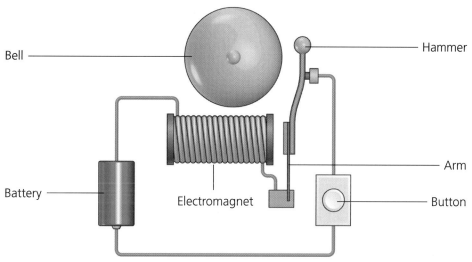

Figure 2
A doorbell uses electricity to make the hammer strike the bell and produce a sound.

TRY THIS: CREATE SOUND WITH ELECTRICITY

Skills Focus: inferring, communicating

Use modelling clay to make a large paper clip stand upright on your desk. Rub a strip of stiff, clear plastic (about 3 cm by 10 cm) several times in one direction with a piece of wool. Quickly place the plastic above the paper clip, and listen!

Now charge a balloon by rubbing it on a piece of wool or on your hair. Quickly bring the balloon near the metal chalk rail of a blackboard, and listen.

1. What did you hear? Describe the sounds you heard.

2. Use what you have learned about sound waves and the movement of electric charges to explain what you heard.

⫸ CHECK YOUR UNDERSTANDING

1. Explain how vibrations and sound are related.

2. Why do you hear thunder after lightning?

3. Place your fingers against the side of your throat, and say "ahh." Explain what you feel and hear, using what you have learned in this section.

Solve a Problem

Make a Telegraph System

People haven't always been able to communicate with each other with e-mail or a telephone. Over 100 years ago in Canada, people used a telegraph system to send messages across vast distances using a system of dots and dashes called Morse code. This system involved using electricity to send messages.

Problem

You want to make a telegraph system to communicate with a partner.

Task

Using only the materials provided, make a working telegraph system.

Criteria

To be successful, your telegraph system must

- be made of a complete circuit connected to a buzzer
- be able to make a buzzer sound when you press down on a switch
- be able to send a message to a partner

Plan and Test

Materials

- 2 pieces of non-insulated wire (1 m long)
- 2 wooden craft sticks
- rubber band
- round toothpick
- electric buzzer or light bulb in a bulb holder
- insulated wire
- cell (battery) in a holder

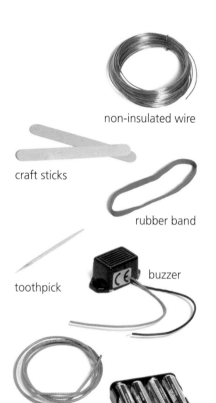

non-insulated wire

craft sticks

rubber band

toothpick

buzzer

insulated wire

battery

Procedure

1. Build a switch that will activate your buzzer. To do this, wrap a piece of non-insulated wire around one end of a wooden craft stick. Wrap the end of the other craft stick with another piece of non-insulated wire. Make sure that you leave enough wire on each end to make the connections. Put the sticks on top of each other so that the wires are touching. Wrap a rubber band around the other end as shown. Insert a round toothpick between the two sticks.

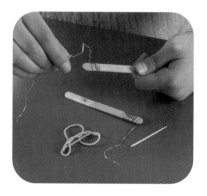

2. Build a circuit using a buzzer or a light bulb, insulated wire, a battery, and your switch.

3. Close the circuit by pressing down on the switch. If your buzzer does not work properly, ask yourself the following questions:
   - Are all the wires firmly attached?
   - Do the wires on the legs of the switch touch when the switch is pressed?
   - Is your buzzer moving on your desk and causing the circuit to break?

Evaluate

4. Use Morse code (**Figure 1**) to send a one word message to a partner. Hold the switch down for a short time to send a "dot." Hold the switch down longer to send a "dash." Pause between the letters in the word.

5. How difficult was it for your partner to distinguish between the sound of a dash and a dot?

Letter	Code	Letter	Code	Letter	Code
a	● ▬	j	● ▬ ▬ ▬	s	● ● ●
b	▬ ● ● ●	k	▬ ● ▬	t	▬
c	▬ ● ▬ ●	l	● ▬ ● ●	u	● ● ▬
d	▬ ● ●	m	▬ ▬	v	● ● ● ▬
e	●	n	▬ ●	w	● ▬ ▬
f	● ● ▬ ●	o	▬ ▬ ▬	x	▬ ● ● ▬
g	▬ ▬ ●	p	● ▬ ▬ ●	y	▬ ● ▬ ▬
h	● ● ● ●	q	▬ ▬ ● ▬	z	● ● ▬ ▬
i	● ●	r	● ▬ ●		

Figure 1
Morse code symbols

Communicate

6. Write an explanation of how electricity is used to make sound.

ⅠⅠ▶ CHECK YOUR UNDERSTANDING ⊗

1. How did you use what you have learned about circuits to build your buzzer?

Chapter Review

Electricity can be changed into other forms of energy.

Key Idea: **Electricity can be transformed into other forms of energy, such as light, heat, motion, and sound energy.**

Vocabulary

light bulb p. 121

Electricity flowing through a light bulb produces light and heat.

An electric motor converts electricity into motion.

Electricity produces sound in a CD player.

Key Idea: **Electric currents produce magnetic fields.**

Vocabulary

magnetism p. 123

electromagnetism p. 123

Electricity flowing through a coiled wire produces an electromagnet.

Review Key Ideas and Vocabulary

When answering the questions, remember to use the chapter vocabulary.

1. What does it mean to transform electricity? Why is transforming electricity important?

2. What is electricity transformed into in each appliance listed below?
   - electric kettle
   - clock radio
   - vacuum cleaner
   - flashlight

3. Why would an electromagnet be used in a recycling yard, rather than a strong permanent magnet?

Use What You've Learned

4. Suppose that you have made an electromagnet, but it is only able to pick up five paper clips. What can you do to increase its strength?

5. In each of the following cases, which experimental set up will produce the stronger magnetic field? Explain your answer in each case.
   - one battery with the wire coiled 20 times around a 5-cm iron nail, or two batteries with the wire coiled 20 times around a 5-cm iron nail
   - two batteries with the wire coiled 30 times around a wooden cylinder, or two batteries with the wire coiled 15 times around a wooden cylinder
   - one battery with the wire coiled 10 times around an 8-cm iron nail, or one battery with the wire coiled 10 times around an 8-cm aluminum nail

6. The switch shown below was just closed. Use what you have learned about electromagnetism to explain what will happen to the nail.

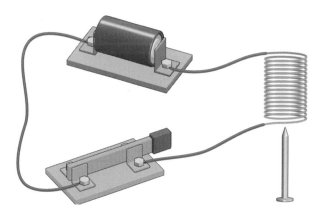

7. What things should an electrician check if the motor of a vacuum cleaner will not turn on?

8. Use what you have learned about light and heat to create an advertisement promoting the use of fluorescent light bulbs or LEDs.

9. Michael Faraday (1791–1867) was a key person in understanding electromagnetism. Find out what he did and what he invented.

www·science·nelson·com

Think Critically

10. How has the electric motor changed the way people work?

Reflect on Your Learning

11. How has learning about energy transformations helped you appreciate the appliances you use?

12. Learning how things work is interesting. Choose an appliance and research how it works.

www·science·nelson·com

Electricity is produced from renewable and non-renewable resources.

KEY IDEAS

▸ Different sources of energy can be used to produce electricity.

▸ Most electricity in British Columbia is produced using hydroelectric energy.

▸ Some sources of electricity are non-renewable.

▸ Some sources of electricity are renewable.

▸ Chemicals can be used to generate electricity.

Electricity is produced from different energy sources, including water and coal. Some energy sources, such as water, are constantly being replaced. Other sources, such as coal, cannot be replaced.

As our demands for electricity increase, there is the possibility that electricity shortages or power blackouts may occur. For example, in August 2003, a massive blackout occurred in Ontario, including Toronto as shown above, and parts of the northeastern United States. Making the best use of the energy sources we have is the only way that we can ensure a plentiful supply of electricity for generations to come.

In this chapter, you will look at how different energy sources are used to generate electricity. You will also look at the production and transmission of electricity in British Columbia.

▷ **LEARNING TIP**

Before you read this chapter, make a web of what you already know about renewable and non-renewable sources of energy.

How Is Electricity Generated?

TRY THIS: GENERATE AN ELECTRIC CURRENT

Skills Focus: creating models, observing

You can generate electricity. All you need are two bar magnets, a piece of copper wire, two leads, and a meter for detecting current. Connect the copper wire to two leads (one at each end of the wire). Connect the ends of the leads to a multimeter. Set up the magnets so that the opposite poles are facing. Hold the wire between the magnets as shown in **Figure 1**. Look at the meter. Now quickly move the wire up and down through the gap.

1. What do you notice on the meter? What does this mean?

Figure 1

In the 1820s, Michael Faraday produced a current by pushing a magnet in and out of a coil of wire. Unfortunately, he had trouble convincing people that his discovery was useful. Eventually, the generator was developed based on Faraday's discovery. Today, we use the effect of a moving magnet on a coil of wire, or a moving wire on a magnet, to generate electricity (**Figure 2**).

The electricity that you use in your home comes from different energy sources. Some energy sources, like water, wind, and energy from the Sun, are **renewable** resources. This means that these sources are constantly being replaced and are always there to use. Other energy sources, like fossil fuels, are **non-renewable** resources. Once non-renewable resources are used up, they cannot be replaced.

Different energy sources are used to turn a turbine [TER-bine], which is attached to a generator. As the turbine turns, the generator produces electricity. In a hydroelectric generator, falling water turns the turbine. Fossils fuels, nuclear energy, wind, the Sun, and even tides also turn a turbine.

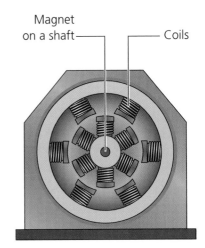

Magnet on a shaft — Coils

Figure 2
Most power stations generate electricity by turning giant electromagnets within stationary coils of wire.

⊪▷ CHECK YOUR UNDERSTANDING

1. What are some energy sources that are used to generate electricity?
2. What energy sources are used to generate electricity where you live?

8.2 How Does Electricity Get to Your Home?

Figure 1
Transmission lines, supported by tall metal towers, carry high voltage electricity over long distances.

Once electricity has been generated, how does it get to you? When the electricity leaves the generator of a power plant, devices called transformers change it to a much higher voltage. The increase in voltage means that less electricity is lost as the current flows along the transmission lines.

The electricity is carried by transmission lines to your community (**Figure 1**). It is then transformed to a low voltage that is suitable to be used in homes. It is transmitted along the street to your home through transmission lines on poles. The voltage of the electricity flowing through your home is at least 110 V, and can be as high as 220 V. This voltage is dangerous, so you should be careful when using plugs and electrical outlets. You should also be careful around transmission lines. People have been **electrocuted,** or killed with electricity, when ladders or equipment they were using touched transmission lines.

The company that provides your electricity needs to know your **consumption,** or how much electricity your family uses. To do this, the company puts a meter on the line that comes into your home (**Figure 2**). Each time you turn on a television or use a hair dryer, the meter records how much electricity you have used. This way, the company knows how much to charge your family.

Figure 2
You can measure how much electricity your family uses by reading dials on the household electrical meter.

Skills Focus: observing, communicating

How does your family use electricity? Observe your family's electrical consumption habits for two or three days. Look at how you and your family are conserving electricity. For example, you may use a hand-held toothbrush rather than an electric toothbrush (**Figure 3**). Your family may use a clothesline to dry clothes or use energy-efficient light bulbs in lamps.

Also look at ways that electricity is being wasted in your home, such as lights being left on in empty rooms.

1. Write down your observations for both conserving electricity and wasting electricity.

2. Create a poster that describes, in words and pictures, different ways to conserve electricity.

Figure 3

The amount of electricity that your family uses can change from day to day. For example, in the winter, when it gets dark early, you might need more electricity for lighting. In the summer, when it is very hot, you might use a lot of electricity to run a fan or air conditioner.

To reduce the amount of money that your family spends on electricity, you should use electricity as carefully and responsibly as possible. As a society, we need to use the available sources of energy to produce electricity carefully and responsibly. This is what energy **conservation** means—taking care of our energy resources in a knowledgeable and responsible way.

⫸ CHECK YOUR UNDERSTANDING

1. How does electricity get to your home? What happens to it along the way?

2. How can you reduce the amount of electricity that you and your family use?

Generating Electricity from Water

In British Columbia, the most commonly used source for generating electricity is moving water. About 85% of all British Columbia's electricity is generated using hydroelectric energy, or **hydro.** Water is a renewable resource because it is constantly recycled on Earth.

Hydroelectric dams, like the one shown in **Figure 1**, are used to change the energy of moving water into electricity. These plants are usually built on large rivers. British Columbians are fortunate to have many fast-flowing rivers that can be used to generate electricity. The first hydroelectric dam in British Columbia was built in 1897 at Bonnington Falls, near Nelson.

Figure 1
Brilliant Dam, near Castlegar, British Columbia, is located on the Kootenay River.

Figure 2 shows a cross section of a hydroelectric dam. A dam stops the flow of water on a river. This creates a large storage lake, called a reservoir. When electricity is needed, water from the reservoir is allowed to flow through a tube, called a penstock, in the dam. At the bottom of the penstock, the water spins a turbine. The turbine drives the generator, which produces electricity.

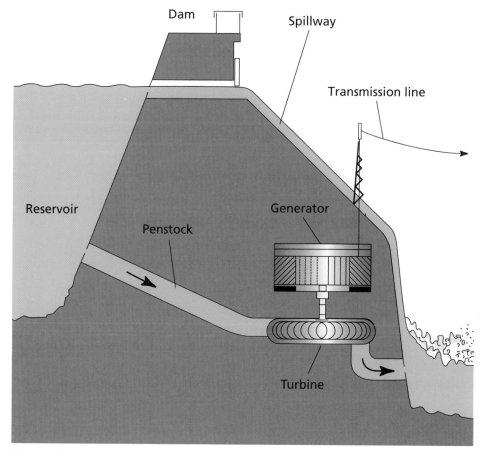

Dam

Spillway

Transmission line

Reservoir

Penstock

Generator

Turbine

LEARNING TIP ◁

As you read about how a hydroelectric dam works, use the labels on the diagram in **Figure 2** to follow the explanation.

Figure 2
The electricity that is generated at a dam is sent along transmission lines to businesses and homes.

Hydroelectric energy is generated without polluting the air. However, the generation of hydroelectric energy creates other environmental problems. For example, large amounts of land are flooded to create a reservoir. The dam itself changes the natural flow of the water. The result is a loss of habitat for many animals and plants, both upstream and downstream from the dam.

The building of dams is a major concern for many Aboriginal peoples in Canada. Dams affect traditional hunting and migration routes. Some hydroelectric energy companies realize the importance of working with Aboriginal peoples and have invited Aboriginal representatives to provide advice. Since many Aboriginal peoples have lived for generations in particular areas, they know a lot about the local ecosystems. This means that they understand the impact of hydroelectric dams on the environment. Some Aboriginal peoples actively protest the construction of dams to protect their traditional lands.

The building of dams is also a concern for the fishing industry. Dams may present a barrier to migrating fish, such as salmon. Fish ladders, like the one shown in **Figure 3**, can be built to help salmon get around dams. Some people feel that this is not enough, however, and that the fishing industry is being damaged by the building of dams.

Figure 3
This fish ladder allows fish to get around dams.

TRY THIS: *MAKE A MODEL OF A WATER WHEEL*

Skills Focus: creating models, communicating

Water wheels use flowing water to do work. The first water wheels were used to grind grain. Today, water wheels called turbines are used to produce electricity. Can you make a working model of a water wheel using common materials? **Figure 4** shows a model that was made with plastic spoons, a Styrofoam ball, and a wooden skewer. To make your model, consider using paper cups, empty creamer containers, wire, foil plates, egg cartons, thread spools, plastic straws, string, and elastics. Pour water over your model to test how it works.

Compare your model with the dam in **Figure 2**. How is your model similar to a dam?

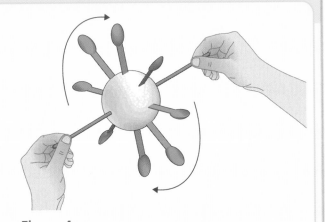

Figure 4
Your model of a water wheel could look like this.

▐▶ CHECK YOUR UNDERSTANDING

1. Why does most of British Columbia's electricity come from hydroelectric energy?
2. What does BC Hydro do? Why is "BC Hydro" a good name for this company?
3. Why is water considered to be a renewable resource?

Producing Electricity from Non-renewable Resources

8.4

There are two types of non-renewable sources of energy: fossil fuels and nuclear [NOO-klee-uhr] fuels. **Fossil fuels** are the most common non-renewable resource used to produce electricity. Fossil fuels are made from the remains of dead organisms that lived millions of years ago. There are three kinds of fossil fuels: coal, natural gas, and oil.

LEARNING TIP ◁

Scan the subheadings in this section. How many types of non-renewable sources of energy do you think you will learn about?

Electricity from Fossil Fuels

In a fossil-fuel generating station, the fuel—either coal, natural gas, or oil—is used to heat water to produce steam. The steam is then used to turn turbines in a generator to produce electricity (**Figure 1**).

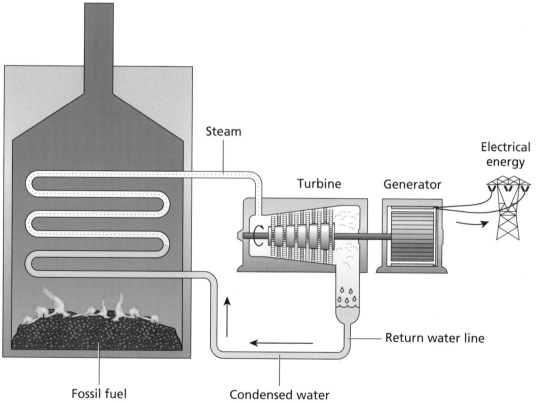

Figure 1
A cross section of a fossil-fuel generating station

Coal is a hard fossil fuel, made of ancient plants such as trees and ferns. Most of the coal that is used in Canada is mined from coal deposits in Saskatchewan, Alberta, and British Columbia.

Natural gas and oil come from plankton (tiny plants and animals) that lived in ancient seas and lakes. Natural gas and oil are usually found together, often in deep wells. Most of Canada's natural gas and oil are found in western Canada. British Columbia has a very large reserve of natural gas and oil that has not yet been developed. Oil wells pump the deposits of natural gas and oil to the surface. Pipelines are used to deliver the fuels to storage basins and refineries (**Figure 2**).

Figure 2
Pipelines are used to transport natural gas and oil. Some pipelines interfere with the annual migration of caribou.

There are many problems with using fossil fuels to create electricity. One problem is that once the fossil fuels are used, they are gone forever. Another problem is that burning fossil fuels puts dangerous gases and particles into the air. As well, there are concerns about coal mining damaging the surrounding environment, and the possibility of oil spills destroying ecosystems.

Electricity from Nuclear Fuel

Most nuclear generating stations use uranium [yoo-RAY-knee-um] as a fuel to create **nuclear energy.** As in a fossil-fuel generating station, the fuel heats water and produces steam. The steam turns a turbine to produce electricity. Although there are no nuclear generating stations in British Columbia, just over 10% of Canada's supply of electricity is provided by nuclear energy. **Figure 3** shows a nuclear generating station in Ontario.

Figure 3
The Pickering Nuclear Generating Station, in Ontario, is one of the largest nuclear generating stations in the world.

Nuclear generating stations are very efficient producers of electricity, and they do not create air pollution. However, many people feel that nuclear energy is not safe. In other parts of the world, accidents at nuclear generating stations have occurred. Harmful radiation [RAY-dee-AY-shun] has leaked and caused serious damage and death. Uranium is radioactive and must be handled very carefully. After it has been used, the waste remains dangerously radioactive for thousands of years. It must be stored carefully, where it cannot leak and pollute the land or air.

⫸ CHECK YOUR UNDERSTANDING

1. What are non-renewable resources? Give examples.
2. Why are fossil fuels considered to be non-renewable resources?
3. What are some of the drawbacks associated with using coal to produce electricity?
4. What are some of the drawbacks associated with using uranium to produce electricity?

8.5 Producing Electricity from Renewable Resources

▷ **LEARNING TIP**

Check for understanding as you read. Turn each of the headings in this section into a question and see if you can answer it.

There are many renewable resources—other than water—that can be used to produce electricity. These renewable resources include sunlight, steam from underground, wind, tides, and biomass.

Electricity from the Sun

You have probably seen or used a solar-powered calculator. Have you ever seen a solar-powered car, like the one shown in **Figure 1**? Both of these devices have solar cells, or photoelectric cells, that use sunlight to produce a small amount of electricity. Energy from the Sun is called **solar energy.** Solar energy can also be used to produce larger amounts of electricity. At a solar-powered station, mirrors are used to focus sunlight on water tanks and heat the water. The steam that is produced turns the turbines to generate electricity.

Figure 1
This car uses solar cells to change solar energy into electricity.

The major benefit to using solar energy is that the supply seems endless to us. As well, solar energy does not create any pollution. However, the amount of sunlight varies with the time of day, the season, and the weather. It is very hard to use solar energy on a dark, stormy day! Solar-powered stations need not only a lot of sunlight, but also a lot of land on which to set up solar collectors and mirrors.

Electricity from Earth

Deep inside Earth is a large amount of energy. This energy is called **geothermal energy.** Geothermal [JEE-oh-THUR-muhl] energy heats water and produces steam. The hot water and steam either reach the surface through cracks or are brought to the surface by drilling.

Geothermal energy can be used to produce electricity. The hot water or steam is used to turn turbines to produce electricity. A power plant that uses hot water or steam is called a geothermal power plant.

There are advantages to using geothermal energy. It does not create pollution, and it is a reliable source of energy. The disadvantage to geothermal energy is that only certain places on Earth have the right geology for this energy. Fortunately, many areas of British Columbia have the right geology (**Figure 2**).

LEARNING TIP ◁

The word "geothermal" combines the Greek word "geo," which means "Earth" with the Greek word "thermal," which means "producing heat or warmth." So "geothermal" refers to heat from Earth.

Figure 2
The South Meager geothermal area is located 170 km north of Vancouver. Research indicates that this geothermal project has the potential to supply 80 000 homes with electricity.

Electricity from the Wind

Wind is another source of energy. **Wind power** has been used for centuries to sail boats and to pump water using windmills. Wind energy can also be used to produce electricity (**Figure 3**). Wind pushes against the blades of a wind turbine and turns them. This turns a magnet that generates electricity.

Figure 3
Wind turbines can change the energy in moving air into electricity.

Several sites are being studied for future wind farms in British Columbia. Most of these sites are located on Vancouver Island and on the northern costal areas of the mainland. In these areas, strong wind flows come off the Pacific Ocean. Most wind turbines need a wind speed of 20 km/h to be useful.

The wind is a source of clean, renewable energy. Because the wind does not blow steadily, however, it is not a reliable source of electricity. As well, some people think that wind turbines are unattractive, noisy, and dangerous to flying birds.

Skills Focus: creating models, communicating

A windmill needs steady winds of at least 11 km/h to work. An anemometer [AN-uh-MOM-ih-tuhr] is a device that is used to find out the speed of the wind. Design a simple anemometer to determine the wind patterns at your school. Your anemometer could look something like the anemometer shown in **Figure 4**. Place your anemometer in different areas around your schoolyard, and at different times.

1. What time of day does the wind blow the hardest?
2. Where in your schoolyard is the wind the strongest? Why do you think this happens?
3. Where does the wind blow most consistently? Why do you think this happens?
4. Where in your schoolyard would you build a windmill? Explain why you would choose that location.

Figure 4

Electricity from the Tides

One very powerful renewable source of energy is the moving ocean. The problem is capturing this energy. One way of capturing this energy involves filling a reservoir with ocean water at high tide and later releasing the water through hydroelectric turbines as the tide goes back out. This is called **tidal energy.** There are only a few tidal power plants around the world. The second largest tidal power plant in the world is the Annapolis Tidal Generation Station in Nova Scotia (**Figure 5**). British Columbia has a number of locations on the coast that are ideal for using tidal energy to produce electricity.

Figure 5
This tidal power plant near Annapolis Royal, Nova Scotia, uses the power of the tides to produce electricity.

While tidal energy is widely available and does not create pollution, it does have drawbacks. Like geothermal energy, tidal energy is only available in certain places on Earth. As well, many people are concerned about how tidal power plants affect the environment. Tidal power plants can harm coastlines and can also affect marine life. For example, in August 2004, a young humpback whale nicknamed Sluice became trapped in the Annapolis River for two weeks. Sluice had entered the river through the gates next to the tidal plant (**Figure 6**). The power plant was shut down to ensure the safety of the whale.

Figure 6
Sluice, the humpback whale, breaches in the Annapolis River as tourists watch. Sluice lived in the river for two weeks before returning to the ocean.

Energy from Living Things

When you burn wood in a campfire, you are using biomass to create heat. The term **biomass** refers to any type of plant or animal tissue, such as wood and wood chips, straw, manure, and crop waste. Biomass can be burned to heat water and create steam to turn turbines and generate electricity (**Figure 7**).

Figure 7
Williams Lake Power Plant is the largest wood-waste-fuelled electric generating plant in North America.

The wood waste from sawmills is the most common source of biomass in British Columbia. In fact, did you know that most pulp and paper mills produce much of the energy they need by burning wood waste?

Although there is a large supply of biomass in Canada, little is used as an energy source. This is because the burning of biomass produces pollution. Scientists are currently developing new non-polluting ways to generate electricity from biomass.

⏵ CHECK YOUR UNDERSTANDING

1. Explain why each source of energy discussed in this section is considered to be renewable.
2. Which renewable energy sources could be used in British Columbia? Explain your answer.

▷ **LEARNING TIP**

Before you begin this activity, read the section "Exploring an Issue" in the Skills Handbook.

Choosing an Energy Source

Electricity always comes at a cost. The most obvious cost is the bill that we pay each month for the electricity we use. There are also costs involved in producing electricity and getting it to our homes. For example, there are costs to the environment, regardless of whether the energy source is renewable or non-renewable. Some energy resources are expensive to obtain and transport. Others create pollution and cause health problems.

The Issue

It is important for communities to consider both the costs and the benefits of different energy sources when choosing an energy source to produce electricity.

Background to the Issue

Electricity can be generated using water (hydroelectricity), fossil fuels, nuclear energy, solar energy, geothermal energy, wind, tidal energy, and biomass. Fuel cells can also be used to generate electricity (**Figure 1**). Some energy sources create pollution or harm the environment (**Figure 2**). Other energy sources are not very practical because suitable technology has not yet been developed or because the resource is only available in certain areas. Which energy source should your community use to generate electricity?

Figure 1
Hydrogen-powered fuel cells were used to light up Vancouver's Science World geodesic dome and welcome the 2010 International Olympic Committee. The fuel cell uses hydrogen and oxygen to produce electricity. The fuel cell is an alternative source of electricity.

Identify Perspectives

To decide which energy source should be used, you must look at the various energy sources from different perspectives. This means that you must look at both the benefits and the drawbacks of each energy source.

Working in groups of three, decide on two or three criteria that you feel are most important when considering the best energy source. These criteria could be cost, the availability of the source, the safety of the source, its impact of the environment, its impact on local Aboriginal groups and their traditional ways of life, and the available technology.

Figure 2
Burning fossil fuels to produce electricity also produces pollution.

Gather Information

Use information presented in this textbook, the Internet, and the library to research three energy sources. Look at each energy source using your criteria. Make sure that you look at both the benefits and the drawbacks of the energy source. Based on your research, can you conclude which energy source should be used in your community?

www.science.nelson.com GO

Make a Decision

Develop a system to rank each energy source according to your criteria. For example, you could rank a very good energy source as 4, a good source as 3, a fair source as 2, and a poor source as 1. Make sure that you can explain your rankings. Add up the points for each energy source. The source with the highest total points will be your choice. If two or more sources tie, decide how you will choose the best. You could look at another criterion to help you make your choice.

Communicate Your Decision

Share what you have learned with your class. You may want to create a poster promoting the energy source you have chosen.

CHECK YOUR UNDERSTANDING

1. You ranked three different energy sources to help you decide which energy source would be the best for your community. Were you surprised to see which source had the most points? Do you think ranking like this helped you to make a good choice? Why or why not?

8.7 Producing Electricity from Chemicals

▷ **LEARNING TIP**

As you read this section, think about your use of batteries. How many batteries do you use? What type of batteries do you use? How do you dispose of the batteries you use?

What devices do you own that run on batteries? We rely on batteries to run many of our electrical devices. A battery creates an electric current from a chemical reaction. Energy can be stored in batteries until it is needed to power electrical devices. As you discovered when you created circuits, batteries only conduct electricity when they are connected to a complete circuit. The best thing about batteries is that they can be taken anywhere (**Figure 1**).

Figure 1
Batteries are convenient sources of electricity.

Alessandro Volta (1745–1827) developed the first electric cell in the 1790s. The cell was made of copper and zinc discs, separated by a thick paper disc that had been soaked in salt water. The zinc and copper discs acted as the electrodes, and the soaked paper disc acted as the electrolyte.

There are different types of batteries (**Figure 2**), but they all work by transforming chemical energy into electricity. Rechargeable batteries can be recharged and used again. A disposable battery is a one-use battery. When the stored energy is used up, the battery cannot be used again. Rechargeable batteries are more expensive than disposable batteries because they cost more to make.

Batteries are convenient, but they are not problem free. Some batteries contain dangerous materials, such as acids and poisonous metals, including lead and mercury. These batteries are usually thrown away and end up in landfill sites. Over time, the dangerous materials can leak into the ground and contaminate it. Also, some materials that are used to make batteries are quite rare. They are found only in specific deposits on Earth.

Today, some manufacturers produce batteries that contain less poisonous metals. Many communities encourage people to take used batteries to recycling facilities that dispose of the batteries safely. Unfortunately, only a very small percentage of batteries are being recycled. Can you think of a way that you could use batteries more responsibly?

Figure 2
There are many types of batteries.

Ⅲ▶ CHECK YOUR UNDERSTANDING

1. What is the difference between a rechargeable battery and a disposable battery?

2. Why is it important to recycle batteries properly or to use rechargeable batteries? Think of ways that you could encourage people to recycle batteries.

3. Copy and complete the following chart in your notebook. Is it better to buy an expensive rechargeable battery or a cheaper disposable battery?

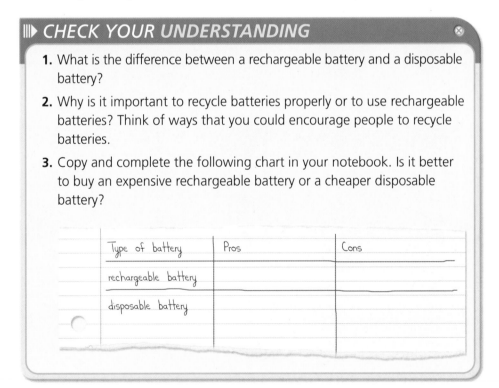

Type of battery	Pros	Cons
rechargeable battery		
disposable battery		

8.7 Producing Electricity from Chemicals

The Human Power House

What is always available and can be used to generate electricity? The answer is you!

When you walk or run, you generate electrical energy. Scientists have developed a way to harness this energy to power a location-sensing mechanism that is mounted on running shoes. As you walk, your heel compresses a crystal, which transforms the force of your step—the mechanical energy—into electricity to power the device. While wearing this device, you never need to worry about getting lost!

Do you have a watch? While most watches are battery operated, some watches are powered by the energy of your swinging arm. As you swing your arm, a device in the watch converts the mechanical energy into electrical energy to power the watch. Does this mean that you have to keep moving your arm to operate the watch? Fortunately, the mechanism also includes storage so that the watch will continue to keep time even when your arm is not moving.

Going for a run or a brisk walk can generate a surprising amount of electricity. But most of that energy is consumed by your body to keep it going. The remaining energy can be used to power small devices, shown on these shoes.

A small weight inside a watch powers a mini-generator. The mini-generator transforms mechanical energy into electrical energy, even while you sleep.

Other watches work by changing body heat into electrical energy. The electricity is produced by a difference in temperature. The heat from the surface of the body is absorbed through the back of the watch, and the temperature of the surrounding air is registered on the face of the watch. The difference in the temperatures is enough to power the watch.

Imagine the amount of electricity that might be produced if the same technology was taken into space or to a volcano! There is a huge temperature difference between the surface of a spacecraft facing the Sun and the surface in the shadow, or the surface and interior of a volcano. These are just two of the applications that scientists are looking into.

This is just the beginning. While most of these technologies are not able to produce very much electricity, they are the first step toward harnessing the potential energy of the human body. What do you think might be next?

Chapter Review

Electricity is produced from renewable and non-renewable resources.

Key Idea: Different sources of energy can be used to produce electricity.

Key Idea: Most electricity in British Columbia is produced using hydroelectric energy.

Vocabulary
electrocute p. 138
consumption p. 138
conservation p. 139
hydro p. 140
hydroelectric dams p. 140

Key Idea: Some sources of electricity are non-renewable.

Vocabulary
non-renewable p. 137
fossil fuels p. 143
coal p. 143
natural gas p. 144
nuclear energy p. 145

Key Idea: Some sources of electricity are renewable.

Solar energy

Geothermal energy

Tidal energy

Biomass

Vocabulary
renewable p. 137
solar energy p. 146
geothermal energy p. 147
wind power p. 148
tidal energy p. 149
biomass p. 151

Key Idea: Chemicals can be used to generate electricity.

Review Key Ideas and Vocabulary

When answering the questions, remember to use the chapter vocabulary.

1. What are some energy sources that can be used to turn the turbines in a generator?

2. What is the main energy source used to generate electricity in British Columbia? Why is it the main source used?

3. List two non-renewable energy sources and explain why each is considered to be non-renewable.

4. List five renewable energy sources and explain why each is considered renewable.

5. Explain how chemicals can be used to generate electricity.

Use What You've Learned

6. Why can hydroelectric energy be used to power a city, while batteries cannot?

7. Some people believe that all nuclear power plants should be shut down and no more should be built. Other people view nuclear energy as a cleaner option to burning traditional fuels, such as coal and oil. Give one reason for each point of view.

8. What are two problems associated with the use of batteries? Provide one possible solution for each problem.

9. Why are availability and reliability important when choosing an energy source?

10. Some electrical sources, such as hydroelectricity, have few environmental costs. Use what you know about the cost of producing electricity to explain why hydroelectricity is not pollution free.

11. Contact your electrical provider. How is most of the electricity generated in your community? What are the environmental concerns for your community? What alternate sources of electrical energy are being considered for your community?

12. British Columbia has not yet developed its offshore oil and gas reserves. Offshore development would mean drilling and building wells off the coast of British Columbia. Find out where the province has reserves of offshore oil and gas. Research some of the environmental concerns about offshore drilling.

www·science·nelson·com **GO**

13. Why is it important to continue developing more efficient ways to produce electricity?

Think Critically

14. Prepare a poster or a pamphlet that explains the dangers associated with transmission lines.

Reflect on Your Learning

15. What three important ideas have you learned about the sources of electricity?

Making Connections

Design Your Own Game

Looking Back

Are you ready to use what you have learned about electricity to create a game? Working in groups, you will make a game that consists of a complete circuit and a set of question cards related to what you have learned throughout this unit. It is up to you how you will incorporate these two elements! Once you have created your game, you can trade with other groups to have some fun and show what you have learned.

Demonstrate Your Learning

Part 1: Build a circuit board.

Use what you have learned about the way electricity travels to build a circuit board. Here are some of the questions you should consider:

- What is the purpose of the circuit board?

- Will it signal a correct response with a light or a buzzer?

- Will it be part of the game itself? Perhaps players must successfully navigate along a "wiggle wire," like the one shown in the photo above, before they can answer a question card.

- What kind of circuit will you build: a series circuit or a parallel circuit?
- What materials do you need to build your circuit board?

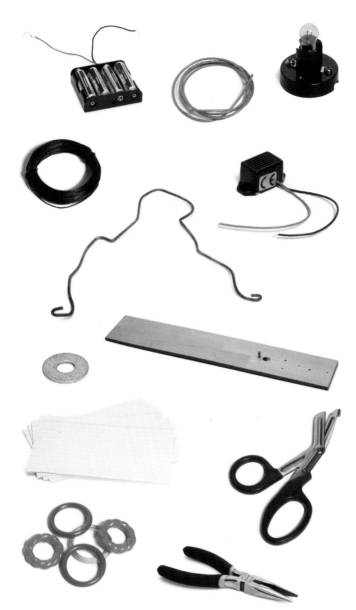

Part 2: Create your game cards.
Create 20 game cards with questions. The game cards must cover all the key ideas you have studied in this unit. Ask a variety of questions based on the information that you have learned. Plan to include about five questions per chapter.

Below are some examples of questions:

- *When* is it better to use a parallel circuit?
- *Why* do some buildings have lightning rods?
- *What* is the connection between rivers in British Columbia and electricity?
- *How* would you find out if a material is an insulator or a conductor?

Write each question on one side of a card. Write a complete answer to the question on the other side of the card. Try not to include questions that can be answered with one word.

Part 3: Plan your presentation.
Plan how you will present your game to the class. Your presentation should include instructions on how to play the game and how players score points or win the game.

⏵ ASSESSMENT ⊗

Your game should provide evidence that you are able to

CIRCUIT BOARD
- build an electric circuit that lights a bulb or sounds a buzzer
- work cooperatively in a group

GAME CARDS
- turn key ideas into questions
- answer questions accurately
- use appropriate scientific terminology

PRESENTATION
- work cooperatively in a group
- communicate clearly
- use appropriate scientific terminology

UNIT C

EXPLORING EXTREME ENVIRONMENTS

CHAPTER **9** Extreme environments are places where human survival is difficult or impossible.

CHAPTER **10** Technology allows us to explore Earth's extreme environments.

CHAPTER **11** Technology allows us to explore the extreme environment of space.

CHAPTER **12** Exploring extreme environments has both benefits and costs.

Preview

Imagine giant plumes of smoke-like clouds and volcanic ash, towering "black smoker" chimneys spewing black water, and clumps of strange-looking tube worms. These were just three of the spectacles witnessed by two Canadian scientists who were part of an expedition to a chain of underwater volcanoes, called the Ring of Fire, in the Pacific Ocean. The international expedition used a Canadian-made submersible to explore the marvels of deep ocean volcanoes. What they discovered astounded them—spectacular creatures and sights that had never been seen before, including an underwater volcano erupting. According to one of the scientists, "We all sat there for five minutes saying 'Wow, this is really cool,' until someone said, 'I think it's erupting!' When they called this trip an exploration, they weren't kidding."

The depths of the oceans are just one of the extreme environments—on Earth and beyond—that scientists explore. In this unit, you will learn about these environments, what makes them "extreme," and the reasons people explore them. You will learn about the technology that allows people to explore places where survival is difficult or impossible, and the important role of Canadian scientists in developing and testing exploration technology. In addition, you will also do some exploring yourself, using your scientific skills to observe, record your data, and make decisions about what you discover—just the way scientists do.

TRY THIS: **THINK ABOUT EXTREME ENVIRONMENTS**

Skills Focus: questioning, communicating

Working in a group, choose an environment that you think is extreme. The environment could be a polar region, a desert, a volcano, the deep ocean, or outer space.

1. Brainstorm what you already know about this environment and what you would like to know. Use questions like these: What makes survival difficult in this environment? How are scientists able to explore this environment? How do they use technology?

2. Create a concept map of your ideas. Use words or pictures, or both to summarize what you know and what you would like to learn about the environment.

◀ "Black smoker" chimneys on the floor of the South Pacific Ocean.

CHAPTER 9

Extreme environments are places where human survival is difficult or impossible.

▐▶ KEY IDEAS

▸ Living things can inhabit extreme environments.

▸ Extreme environments include polar regions, deserts, oceans, volcanoes, and space.

▸ People explore extreme environments to find resources and to better understand our world.

▸ Technology allows humans to survive in extreme environments.

Earth may be the only place in the universe that can support life, but it is not actually a very welcoming place for humans. Most of the planet is covered with vast oceans. Of the part that is land, large areas are either too cold or too hot for humans to survive easily.

However, people do live in places where survival is difficult, such as deserts and the Arctic. Scientists live in research stations in the extreme conditions of Antarctica, shown above. People have even explored uninhabitable areas, such as volcanoes, oceans, and outer space. Humans have been able to explore and settle in environments far beyond where they could survive unaided because they have developed the technology to overcome the obstacles to survival.

In this chapter, you will learn about extreme environments. You will learn why people explore these environments and how people use technology to survive extreme conditions. You will even get to plan your own extreme expedition!

TRY THIS: DESCRIBE YOUR ENVIRONMENT

Skills Focus: observing, inferring

How hot does it get in summer where you live? How cold does it get in winter? Your answers describe the extremes of temperature in your environment. But what would happen if the temperatures became even more extreme?

1. Describe how hotter or colder temperatures would affect the way you live. Think about the effect on your outdoor activities, the clothing you would need, and the home you live in.

2. What would you and your community have to do to adapt to the changes in temperature?

You are surrounded by your **environment.** Other living organisms, non-living objects, such as soil and buildings, and even the weather all make up your environment. To survive, you, like all living things, must be able to meet your basic needs for water, food, air, and shelter within your environment (**Figure 1**).

Figure 1
Winter can be severe in the northern town of Stewart and other parts of British Columbia. However, people survive extreme temperatures by building structures that withstand the cold and snow, and by wearing protective clothing.

> **LEARNING TIP**

Review this section to identify the main ideas. Can you identify the most important idea? If anything isn't clear to you, reread the section.

Some environments have very harsh conditions. They may be extremely hot or extremely cold, there may be no water, or there may be no air to breathe. These **extreme** conditions make it difficult, and in some cases impossible, for humans to survive. For humans, these are **extreme environments.**

However, not all organisms need the same living conditions as humans do. Living things are found everywhere on Earth—in the hottest deserts, in the coldest oceans, and even deep inside Earth! There are heat-loving bacteria that thrive in hot springs and volcanoes. Many organisms, including fish and giant squid (**Figure 2**), are found in the deep sea. In fact, scientists recently discovered 38 000 new species of ocean life in just four years of research!

Figure 2
Scientists recently discovered a giant squid, nicknamed "Colossal Squid," living deep in the ocean. It weighed over 150 kg and was over 5 m long.

These organisms can survive conditions that are considered extreme for humans because they have adapted to the conditions. Obviously, to these organisms, the conditions are not extreme, but are just right.

||► CHECK YOUR UNDERSTANDING   ⊗

1. What makes an environment extreme?
2. What conditions would make life difficult or impossible for humans to survive?
3. Why can some bacteria live in volcanic vents?

The Extreme Environments of Earth and Beyond

For humans, an extreme environment is a place where the conditions are so harsh that human survival is difficult or impossible. Polar regions, deserts, oceans, volcanoes, and space are examples of extreme environments. Although each environment is different, they are all characterized by extreme conditions, such as very hot or very cold temperatures, little or no water, crushing pressure, or no air.

Figure 1 is a map of the world showing the locations of extreme environments.

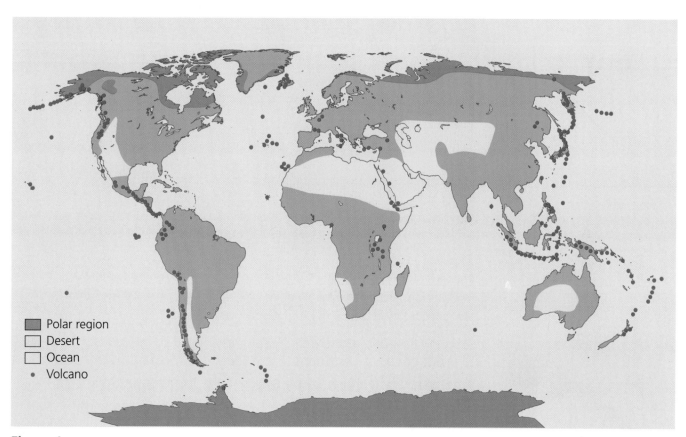

Figure 1

Polar regions (in purple), deserts (in yellow), oceans (blue), and volcanoes (red dots) are extreme environments.

<table>
</table>

> **LEARNING TIP**
>
> Look through the headings in this section. Before you read the section, predict what makes each of these environments extreme.

Polar Regions

There are two polar regions: the Arctic and Antarctica. The Arctic is the cold area around the North Pole. It includes the Arctic Ocean at the centre and the land that surrounds the Arctic Ocean. It is usually identified as the area north of the Arctic Circle. The average annual temperature is 0 °C. The Arctic is home to many animals, such as polar bears, arctic foxes, arctic wolves, walruses, seals, and whales. These animals have developed adaptations to the frigid conditions. People, such as the Inuvialuit [in-oo-vee-AL-oo-it], also live in the Arctic (**Figure 2**). They have survived by developing the technology to protect themselves from the extreme cold.

Figure 2
The town of Inuvik, Northwest Territories, is the largest town north of the Arctic Circle. The extreme minimum temperature is −56.7 °C. Inuvik receives 56 days of total sunlight in summer and 30 days of total darkness in winter.

Antarctica is a continent that is located almost entirely within the Antarctic Circle. It is the coldest, highest, and windiest place on Earth, and holds the record for the coldest temperature: −89.2 °C. Since Antarctica is mostly covered with ice, most of the animals that live on land are microscopic animals and insects. The penguins that live on the coasts are the one exception. In comparison to the land, the ocean surrounding Antarctica supports a wide variety of life, such as zooplankton [ZOE-eh-PLANK-tuhn], penguins, seals, whales, and dolphins. Antarctica is the only continent that has no permanent population of humans. However, scientists from all over the world, including Russia, Japan, the United States, Australia, and New Zealand, work and live in scientific stations in Antarctica.

Deserts

Deserts are very dry areas that get less than 25 cm of rain a year (**Figure 3**). Most people think that deserts are hot places, but deserts can also be very cold. Both the Arctic and Antarctica are deserts. The Atacama desert in Chile is the driest place on Earth. It receives less than 0.01 cm of rain per year. Some deserts are sandy, but most are wildernesses of rock and stone. Animals and plants that live in deserts have special adaptations that allow them to survive, despite the lack of water. People also live in some deserts. For example, the Afar people of Ethiopia, Africa, live in one of the hottest place on Earth, where temperatures are between 35 °C and 40 °C year-round. The living conditions are harsh, but the Afar have found ways to survive.

Figure 3
The "Pocket Desert" near Osoyoos, in the South Okanagan Valley, British Columbia, gets less than 20 cm of rain a year. About one-fifth of Earth's land surface is desert.

Oceans

Most of Earth's surface (about 71%) is covered by oceans. Even though people can travel on the surface of the ocean, the deepest parts of the ocean are very hard to explore. The ocean depths are pitch black. There is no air to breathe, and the crushing pressure increases the deeper you go. **Figure 4** shows some of the amazing creatures that live in the ocean depths.

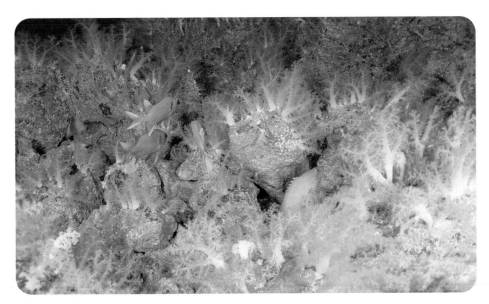

Figure 4
Life near thermal vents in the Mariana Trench, the deepest part of the ocean. This photo was taken by ROPOS, a Canadian robotic submersible.

Volcanoes

Volcanoes are vents in Earth's surface where molten rock from below the surface can rise up and spill over (**Figure 5**). Volcanoes are particularly dangerous because they are unpredictable. Some volcanoes are dormant, or sleeping, and do not erupt. Active volcanoes produce hot lava, steam, and ash, which can destroy the surrounding area. Today, there are over 1000 active volcanoes on land and many more in the oceans. Volcanologists study volcanoes to predict when they may erupt and to understand what causes them to erupt.

Figure 5
Kilauea, in Hawaii, is the world's most active volcano. Lava erupts frequently, which poses a danger to the surrounding area and the people who live there.

Volcanoes are located in specific parts of the world. One of the most important areas that scientists are exploring is a chain of volcanoes that encircles the Pacific Ocean. This chain is known as the Ring of Fire (**Figure 6**). Indonesia has the most active volcanoes. Scientists believe, however, that there are probably more volcanoes on the ocean floor than there are on land.

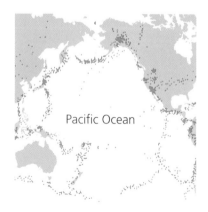

Pacific Ocean

Figure 6
Ring of Fire

Outer Space

For centuries, people have looked up into the sky and imagined travelling through space (**Figure 7**). Space is the region beyond Earth's atmosphere [AT-muhs-FEAR]—the ultimate extreme environment. The temperatures are *really* extreme—from boiling hot in the light of the Sun to freezing cold in the shadow of Earth—and there is no air to breathe. As well, space is close to a vacuum. Space contains atoms and particles of dust, but they are spread so thinly that space is considered to be empty.

Figure 7
People have always wondered about what is in space. However, it is only in the past 50 years that people have been able to travel to and survive the extreme conditions of space.

⫸ CHECK YOUR UNDERSTANDING

1. What are some characteristics of an extreme environment?

2. What is the main characteristic of a desert? Is a desert always a hot place? Explain why or why not.

3. Why do oceans present such a challenge for human exploration?

4. Create a Venn diagram to compare and contrast the challenges of exploring in space with the challenges of exploring the oceans.

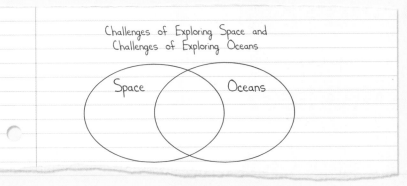

Challenges of Exploring Space and
Challenges of Exploring Oceans

Space   Oceans

5. Choose one extreme environment. Discuss the conditions in this environment that affect human travel.

Why Do People Explore Extreme Environments?

9.3

▷ **LEARNING TIP**

Before you read section 9.3, see if you can answer the question in the title at the top of the page. As you read, check to see if your answer was correct.

Why would anyone want to go to the cold polar regions, hot deserts, deep oceans, dangerous volcanoes, or even space? After all, these are harsh places for humans to live. But as long as humans have existed, they have gone on **explorations,** or voyages into unknown territory to investigate new frontiers and to search for new discoveries. Let's look at some of the reasons people explore extreme environments.

Exploring to Find New Places to Live

As populations grew, ancient peoples explored to find new places to live. They also explored to find food, water, and other resources. Today, people explore to learn about the world and the universe, rather than to find a new place to live. However, some scientists and politicians talk about having a human colony on the Moon or Mars some day.

TRY THIS: LOOK AT KWADAY DAN TS'INCHI

Skills Focus: observing, inferring

Ancient peoples faced danger when they travelled away from their home settlements. In 1999, the frozen remains of a male Aboriginal, from about 500 years ago, were discovered in a glacier in British Columbia. The discovery was named Kwaday Dan Ts'inchi, which means "long ago person found" in the Southern Tutchone language. Scientists and members of the Champagne and Aishihik First Nations are studying the remains and the items that were found with him. These items include a carved walking stick, a knife-like tool in a hide and fur sheath (**Figure 1**), a fur robe, a woven hat, a bead on a thong, and dried salmon.

1. Look at the list of items that were found near the frozen remains of the long ago person. What do you think he was doing? What are some of the dangers he might have faced?

2. Why would dried salmon be a good thing to carry?

3. What can we learn from the man's possessions?

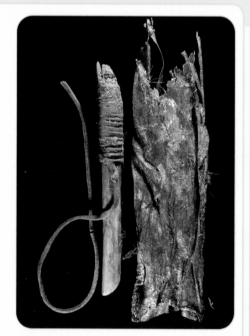

Figure 1
This knife-like tool was found with Kwaday Dan Ts'inchi, along with the hide and fur sheath that held it.

Exploring to Understand Our Past

We explore extreme environments to learn about our history, the history of other living things, and the history of Earth itself. For example, deep-sea divers look at the remains of shipwrecks (**Figure 2**). Paleontologists [PAY-lee-on-TALL-uh-jist] look for fossils of extinct creatures. Archeologists [ARE-key-ALL-uh-jist] study ancient civilizations to understand how people lived in the past. Scientists study the layers of ice in Antarctica to learn how Earth's atmosphere has changed over time.

Figure 2
A diver explores a shipwreck off the coast of Florida.

Exploring to Find New Plants and Animals

When explorers travelled to new lands, they often found plants and animals that were different from the plants and animals at home. Today, we explore extreme environments to discover new species of plants and animals. For example, scientists have discovered new species of corals and sponges on the ocean floor (**Figure 3**). Scientists have also discovered life in volcanic vents, in hydrothermal vents on the ocean floor, and in desert rocks.

Figure 3
Scientists hope that species of sponges and corals, newly discovered on the ocean floor, may be used to develop medicines to treat human diseases.

9.3 Why Do People Explore Extreme Environments?

Exploring to Find Resources

The ocean beds contain vast reserves of fossil fuels, such as natural gas and oil (**Figure 4**). We use fossil fuels to power our cars, heat our homes, and produce electricity. Because the current reserves of fossil fuels are running out, scientists are looking for new reserves. Ocean exploration has led to the discovery of fossil fuel reserves beneath the ocean floor. The Arctic is another environment that is being explored for its fossil fuel reserves.

Figure 4
Semi-submersible oil rigs, like this one in Hibernia, are used to drill for oil in the ocean floor.

Exploring to Understand and Protect Our World

In the past, explorers searched for new lands and easier sea routes. Today, people explore to understand and protect our world. For example, scientists have been able to track elephants moving through the Sahara Desert using satellite images (**Figure 5**). They have learned that these elephants survive the hot, dry desert by moving from one watering hole to the next at just the right time. Scientists can use this knowledge to protect the elephants.

Figure 5
By tracking the movements of these elephants through the desert, scientists can learn how to protect them.

⫸ *CHECK YOUR UNDERSTANDING* ⊗

1. Why did ancient peoples explore?

2. Use a Venn diagram to compare the reasons why ancient peoples explored with the reasons why people explore today.

3. Give an example of how satellite images can help scientists protect endangered animals.

4. Work with a partner to brainstorm a list of explorations you know about from previous learning. In each case, try to identify the reason for the exploration.

9.3 Why Do People Explore Extreme Environments?

9.4 Using Technology to Survive

Humans do not have the adaptations that allow other organisms to survive extreme environments. Instead, humans have developed the tools, or the **technology,** to make it possible to survive in challenging environments. For example, Aboriginal peoples have used technology for thousands of years to survive the challenging conditions in the northern areas of Canada.

Let's take a closer look at some of the ways that the Inuit were able to survive in the extreme conditions of the Arctic.

Snow Houses

Figure 1

The Inuit snow house is an engineering marvel (**Figure 1**). It was constructed in the form of a dome, using blocks of snow. To make sure that the snow was strong enough to be cut into blocks and stacked, the Inuit tested it using a long probe made of antler. A snow house had a fireplace in the centre, surrounded by sleeping platforms. It also had a window to let in sunlight. The floor and living areas were covered with animal skins, making them warm and comfortable. Today, some Inuit still build and live in snow houses for a short time.

Snow Sleds

Figure 2

The best way to travel across ice and snow is to use a komatik [COME-uh-tick], or Inuit sled (**Figure 2**). A komatik could carry an entire household to better hunting and fishing grounds. Traditional komatiks were made of wood and hide, and were pulled by dogs. Modern komatiks are made of metal and plastic, and are pulled by snowmobiles.

Snowshoes

Figure 3

Have you ever tried to walk in deep snow? You probably did not get very far before you sank! Aboriginal peoples invented snowshoes, which allowed them to travel on top of the snow rather than through it (**Figure 3**). Snowshoes spread the wearer's weight over a large area so that the wearer does not sink into the snow. Snowshoes were originally made from bent pieces of wood, with crisscrossed pieces of animal hide. Today, many snowshoes are made using a metal frame.

Parkas

The Inuit understood the way that heat moves and designed their clothing to protect them from very cold temperatures. They made clothing from animal hides and fur. They wore parkas—tight-fitting, double-layered, hooded jackets—to prevent their body heat from escaping into the icy environment (**Figure 4**). Inuit garments were tight around the neck and waist to prevent heat loss, but were open at the bottom. This opening allowed moist air to escape and prevented the wearer from getting sweaty.

Figure 4

Sunglasses

The glare from the Sun, as it reflects on the snow, is very intense. Without protection, human eyes can be damaged, causing temporary "snow blindness." Aboriginal peoples used a piece of bone with a small slit as sunglasses (**Figure 5**). Although low tech compared with the sunglasses used today, these sunglasses reduced the amount of light that entered the eye.

Figure 5

Oil Lamps

Aboriginal peoples of the Arctic have been making oil lamps out of soapstone for at least 2500 years (**Figure 6**). The oil lamp was usually one metre long. It burned oil from melting blubber and used a dried grass wick. The oil lamp had a huge impact on life in the Arctic because it provided light during the winter darkness. It also provided heat for soapstone cooking pots.

Figure 6

⫸ CHECK YOUR UNDERSTANDING ⊗

1. Why was the invention of the komatik important to the exploration of the Arctic?

2. How does wearing snowshoes prevent you from sinking into deep snow?

3. What extremes in the Arctic environment have the Inuit overcome with technology?

9.4 Using Technology to Survive

Figure 1
Michael Schmidt is a scientist who works in remote places, such as the top of Mount Logan in the Yukon Territory. He sets up Global Positioning System (GPS) receivers to study and learn about earthquakes. GPS receivers work with satellites that revolve around Earth. They can locate where something is on Earth, within a few metres.

▷ **LEARNING TIP**

For a review about solving a problem, see the Skills Handbook section "Solving a Problem."

Planning for an Extreme Environment Expedition

Problem

You are part of a team of scientists that will join Dr. Schmidt on a 30-day expedition to Mount Logan (**Figure 1**). Mount Logan is 5959 m high. It is Canada's highest peak and the second highest peak in North America. What will you need to take on your expedition?

Task

Create a list of items you will need for your expedition.

Criteria

You need air, warmth, food, and water. To be successful on your expedition, your list must include

- equipment to help you breathe at the top of the mountain where the air is thinner
- ways to protect yourself from the cold, wet, and wind
- food and water, as well as ways to cook the food
- ways to transport yourself and any equipment you may have
- equipment for communication

Plan and Test

1. You will work in teams of four. Each member of your team will list the items needed for one of the four needs below.

   - *Protection:* What clothes or items will you need to protect you from the cold, wet, and wind? Will you need special breathing equipment to breathe easily on the top of the mountain?
   - *Energy:* What food and water will you need? What equipment will you need for cooking?
   - *Transportation:* How will you carry all your equipment?
   - *Communication:* What equipment will you need for communicating with other scientists and your family?

2. Create a word web like the one below for your portion of the travel plan.

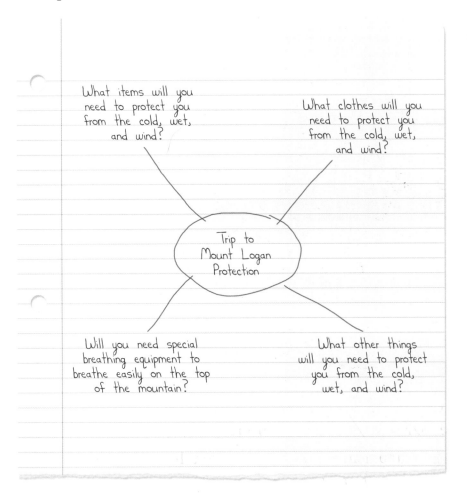

LEARNING TIP ◁

Word webs are a type of concept map. Think about other ways you could organize your ideas about the items you need for your expedition.

3. Use your web to make a list of things you will need.

4. Find a partner on another team who is working on the same topic. Compare your lists. How are your lists similar? How are they different?

Evaluate

5. Together, make a new list that combines the best of your two lists.

Communicate

6. Gather with your original team members. Share your lists (**Figure 2**). Put together all four lists of things needed for the expedition.

Figure 2
This group of students used their lists to develop a dramatization of an expedition to Mount Logan.

⏵ CHECK YOUR UNDERSTANDING

1. Did comparing your list with your partner's help both of you to make a better list? Why or why not?

2. What are the benefits and challenges of working in groups in science? Do you think scientists face the same challenges and experience the same benefits? Why or why not?

3. In this activity, you made decisions about the types of equipment you would need for an expedition. When you make a decision, there is always more than one option you must consider. Why is it important to think about the consequences of each option before making a decision? What can happen if you make a decision without thinking about the consequences?

A Cool Job

How would you drill through thick glacier ice to reach the rock surface below? This was a problem that Dr. Garry Clarke, a glaciologist, and his team had to solve while studying Trapridge Glacier in the Yukon Territory.

To drill a hole in the ice, the team changed a carpet steam-cleaner into a hot-water drill (**Figure 1**). The team placed sensors in the hole to measure how fast the glacier was moving and how it moved over the land. The sensors were connected to data loggers that recorded data. The data loggers were powered by solar panels.

tents, fuel, safety equipment, communication equipment, and all the scientific equipment for their investigations.

Figure 2
This camp is where the members of Dr. Clarke's team cook their food and store their portable computers. They also wash their laundry and hang it to dry.

Figure 1
This hot-water drill can make a 70-m deep hole in the ice in about one hour.

Dr. Clarke's team lived beside the glacier for one month (**Figure 2**). Because the glacier is hundreds of kilometres from the nearest settlement or road, the team needed to bring all their supplies. They brought clothing, food,

Dr. Clarke hopes that looking through a hole in the ice will help him learn what Canada was like during the Ice Age and how our climate is changing today.

These are just some of the things that a glaciologist does. Find out more by doing an Internet search. Share what you learn with your classmates.

www.science.nelson.com

9 *Chapter Review*

Extreme environments are places where human survival is difficult or impossible.

Key Idea: Living things can inhabit extreme environments.

Giant squid live in the deep ocean.

Vocabulary

environment p. 165

extreme p. 166

extreme environments p. 166

Key Idea: Extreme environments include polar regions, deserts, oceans, volcanoes, and space.

Polar regions   Deserts   Oceans   Volcanoes   Space

Key Idea: People explore extreme environments to find resources and to better understand our world.

We explore to find resources.

We can learn to understand the world.

Vocabulary

explorations p. 172

Key Idea: Technology allows humans to survive in extreme environments.

Snow houses and komatiks are just two examples of Aboriginal technology.

Vocabulary

technology p. 176

Review Key Ideas and Vocabulary

When answering the questions, remember to use the chapter vocabulary.

1. Give five examples of living things that inhabit extreme environments.

2. Give an example of an extreme environment, and then explain what makes this environment extreme.

3. Describe all the reasons why people explore extreme environments.

4. List two resources that have been found in extreme environments.

5. Give six examples of technology that enabled Aboriginal peoples to survive in the Arctic.

Use What You've Learned

6. Why was finding the frozen remains of Kwaday Dan Ts'inchi helpful to scientists and members of the Champagne and Aishihik First Nations?

7. Scientists tell us that the temperature in the Arctic has been rising in the past few years. This is causing the sea ice to melt and affecting the lives of the organisms that live in the Arctic, including humans. Imagine that you are part of a Canadian scientific exploration team that is studying the changing Arctic temperature. Work in a group to brainstorm some questions that the exploration team might want to answer.

Think Critically

8. In the past 20 years, what new technology do you think has most drastically changed our world? You may want to use the Internet to research important inventions. Defend your choice.

www.science.nelson.com

9. Why do you think a changing climate would force people to move?

Reflect on Your Learning

10. What are the benefits and challenges of working in collaboration with other scientists, as you did when planning an expedition to Mount Logan?

11. What is the most important idea you learned about extreme environments? Explain why you think this idea is important.

Technology allows us to explore Earth's extreme environments.

▶ KEY IDEAS

- ▶ Technology allows us to navigate Earth's extreme environments.

- ▶ Technology allows us to survive extreme temperatures.

- ▶ Technology allows us to explore the ocean depths.

- ▶ Canadians make important contributions to ocean exploration.

Earth's extreme environments are some of the most fascinating places for humans to explore. But just getting to these places can be difficult and dangerous. Surviving once you're there can be just as challenging, if not impossible.

Humans do not have the adaptations that allow some organisms to survive extreme conditions. Instead, we rely on technology to make possible things that would otherwise be impossible—things like exploring an erupting volcano (as shown in the above photo), living in the frigid cold of Antarctica, or travelling to the deepest ocean depths. In this chapter, you will look at the technology that allows humans to travel to and survive in Earth's extreme environments.

Finding Our Way

Early peoples and explorers used the Sun and stars to guide them on their journeys. For example, the Inuit relied on their detailed knowledge of the night skies and the position of the Sun to help them navigate. After the compass was invented, explorers used it to find out where they were. Today, people use technology, such as radar, satellites, and sonar, to explore and navigate extreme environments.

TRY THIS: USE RADAR TO "SEE"

Skills Focus: creating models, measuring, inferring

In this activity, you will use the concepts of radar to find out the shape of an object in a box. Your teacher will give you a box that has a grid taped to the top. Inside the box is an object. Working in a group, use a nail to poke a hole through the grid. Then use a wooden skewer to probe for the object. When you feel the skewer touch the object, stop and measure the distance, in centimetres, from the top of the box to the end of the stick as shown in **Figure 1**. Record the distance in a table. Repeat this process until you have recorded all the measurements.

Figure 1

1. Create a graph or contour map of your data. If you have access to a computer, you can use a spreadsheet program to create a surface graph.

2. Based on the picture you created, what object is inside the box?

▷ **LEARNING TIP**

Pause after each paragraph on this page and see if you can define the highlighted word in the paragraph.

Radar, which stands for **ra**dio **d**etection **and r**anging, uses radio waves to help people explore. Radio waves are invisible waves that carry voices, music, pictures, and signals through the air. A radar set picks up any echoes that are bounced back off an object and uses the echoes to tell the distance, speed, direction of motion, and shape of the object. Radar systems are used on boats and ships to search for land, ice, and other boats or ships.

Satellites use radar to relay signals for cell phones and television signals. A **satellite** is an object in space that revolves around Earth or any other planet. One of the most important satellite technologies that is used today, for both land and sea navigation, is the Global Positioning System, or GPS. GPS has 24 orbiting satellites that send out radio signals. The boat in **Figure 2** has a GPS receiver that can detect these signals. Signals from three satellites are used to tell exactly where the boat is and how fast the boat is moving. GPS is a good navigation system for environments, such as oceans and deserts, that have few features to use as a reference point. GPS is used in cars, boats, helicopters, ships, submarines, and airplanes, as well as in small handheld receivers that are carried by hikers and explorers.

▷ **LEARNING TIP**

Look closely at **Figure 2**. Then reread the paragraph above it. Now look again at each part of the diagram and check that you understand what it shows about how GPS works.

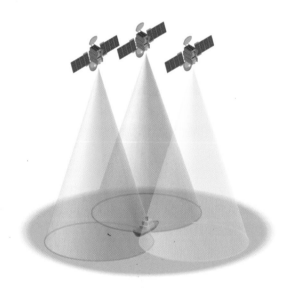

Figure 2
Using radar, GPS satellites can pinpoint the exact location of this boat.

Did you know that ships use sound to chart the depths of the oceans? **Sonar,** which stands for **so**und **na**vigation and **r**anging, works by using the echoes of sound waves. First, a device sends a sound wave into the ocean. The sound wave bounces off the ocean floor. This creates an echo, just like when you shout into a canyon or an empty room. A receiver on the ship can figure out how far away the ocean floor is by measuring the time it takes for the echoes to return to the ship (**Figure 3**).

LEARNING TIP ◄

Look at the highlighted words *sonar* and *radar*. They are formed from the first letters of the words they stand for. Words like these are called acronyms. Other examples in this unit include *scuba* and *NASA*.

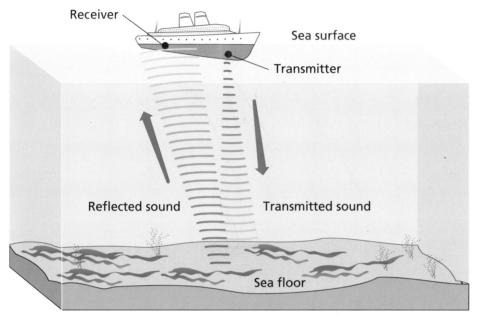

Figure 3

A sonar transmitter underneath a ship sends sound waves down through the water. The time it takes for the sound waves to bounce off the floor and back to a receiver on the ship is used to figure out the depth of the ocean and to create maps of the ocean floor.

Sonar is used to map the ocean floor and to determine the locations of underwater objects—from shipwrecks to submarines. As well, sonar is used to help ships and submarines navigate through shallow and rocky waters.

⫸ CHECK YOUR UNDERSTANDING

1. What does the word "radar" stand for? How does radar work?
2. Compare the GPS system with the navigation systems of the Inuit.
3. Why is GPS such a valuable technology for explorers to use?
4. How is GPS used in everyday life?
5. Bats and dolphins use a technique called echolocation to navigate and to locate prey. They emit sounds and listen for the echoes. Explain how echolocation is similar to sonar.

10.2 Protection from Extreme Temperatures

A polar bear has a nice thick fur coat to keep it warm in extremely cold temperatures (**Figure 1**). But humans do not have natural adaptations to keep them warm. So before humans could travel to and explore very cold or very hot places, they had to develop clothing that would protect them from the extreme temperatures. For example, a winter coat keeps you warm in cold temperatures because it is a good **insulator.** This means that it stops the heat of your body from moving into the surrounding air. Heat is the movement of energy from a warm object to a cool object.

Different materials have different insulating properties. For example, metals are not good insulators because they move heat away from a warm object. Fur is a good insulator because it traps small pockets of air. Trapped air is an excellent insulator because air does not conduct heat very well. Feathers also trap air (**Figure 2**), and so do wool fibres. This is why wool sweaters keep you so warm.

Figure 1
Polar bears keep warm in the cold Arctic temperatures because they have thick fur and a thick layer of fat under their skin. Fur and fat are good insulators because they stop body heat from going into the environment.

Figure 2
The Emperor penguin's feathers provide insulation against the bitter cold of Antarctica. Penguin chicks have fluffy feathers that trap even more air to hold in the heat from the chick's body.

Skills Focus: observing, inferring

Some animals that live in the Arctic have a thick layer of fat, or blubber, under their skin. To see how well blubber insulates, make a "blubber mitt" by filling a large plastic bag about halfway with vegetable shortening or lard (**Figure 3**). Put an empty plastic bag on your hand so that you are wearing it like a mitten. Slide your hand into the shortening-filled bag. Mush around the shortening until it surrounds your hand. Plunge your other hand (your bare hand) into a bucket of ice water. Be careful not to keep your hand in the ice water for too long. Now plunge your blubber mitt into the ice water. What do you notice about your hands?

Figure 3
A "blubber mitt"

▐▶ *CHECK YOUR **UNDERSTANDING***

1. How does insulation keep you warm?
2. Explain how polar bears can survive in the extreme cold of the Arctic.
3. Explain how scientists use observations of nature to develop insulating materials.

SKILLS MENU

○ Questioning   ● Observing
○ Predicting   ● Measuring
● Hypothesizing ○ Classifying
● Designing   ● Inferring
  Experiments
● Controlling   ● Interpreting
  Variables   Data
○ Creating   ● Communicating
  Models

Testing Materials for a Polar Suit

Scientists who work in Vostok, Antarctica, experience the world's lowest temperatures. To survive and work in temperatures as low as −88 °C, scientists need clothing that keeps them warm and dry, and protects them from the wind.

In this investigation, you will test different materials that could be used to make a polar suit for scientists who work in Vostok.

Question

What is the best insulating material for a polar suit?

Hypothesis

Write a hypothesis that states which material(s) would make the best insulator(s). Complete your hypothesis with a short explanation of your reasons. Write your hypothesis in the form "If … then … because … ."

Materials

apron

baby-food jars with cut-out lids

scraps of fabric or other materials

cold water   hot water

hot and cold water in beakers

rubber bands

thermometer

- apron
- 2 or more baby-food jars
- 2 or more lids for baby-food jars, with a hole in the middle large enough for a thermometer to fit
- scraps of fabric or other materials
- hot and cold water
- rubber bands
- thermometer

✋ Thermometers are fragile and expensive. Be careful when handling a thermometer.

Decide what other materials you will need. Check with your teacher to make sure that these materials are safe for you to use.

- Design a procedure to test the insulating properties of different fabrics. A procedure is a step-by-step description of how you will conduct your experiment. It must be clear enough for someone else to follow and do the exact same experiment.

- Submit your procedure (including any safety precautions), to your teacher for approval. Also submit a diagram, at least half a page in size, showing how you will set up your equipment.

Data and Observations

Create a data table to record your observations. Record your observations as you carry out your experiment.

Analysis

1. What did you learn about the insulating properties of different materials?

2. Compile the findings of all the groups in your class. Which materials provided the best insulation?

3. Can you use a graph to represent the findings of your class? Try it and see.

Conclusion

Go back to your hypothesis. Did your observations support, partly support, or not support your hypothesis? Write your conclusion.

Applications

1. Most polar suits contain several layers of different materials. Which combinations of materials do you think would produce an even more effective insulator? If there is time, test your ideas.

2. Would the materials you tested be practical for Arctic explorers (for clothing or protective equipment)? Explain.

LEARNING TIP ◁

For help with your experiment, read the Skills Handbook sections "Designing Your Own Experiment," "Writing a Hypothesis," "Controlling Variables," and "Writing a Lab Report."

▌▐▶ *CHECK YOUR UNDERSTANDING* ⊗

1. Which variable did you change (your independent variable)?

2. How did you measure the change in temperature (your dependent variable)?

3. Were other groups testing the same hypothesis as your group? How were their results the same or different from yours?

Exploring Beneath the Ocean

> **LEARNING TIP**

The first paragraph in this section describes three obstacles to ocean exploration. Turn these into three questions that you should be able to answer after reading the paragraph.

The oceans are the last unexplored place on Earth. Even though most of Earth is covered with water, only a small fraction has been explored. There are many obstacles to ocean exploration. Divers need to be able to breathe underwater and to control their floating and sinking. They also need to be able to overcome the enormous and rapid change in pressure when travelling deep underwater.

The Challenge of Breathing Underwater

The possibility of finding treasure in sunken shipwrecks prompted many inventions that allowed divers to work underwater. In the early 1800s, divers wore a heavy copper helmet attached to a canvas suit. Air was pumped through a long hose that was connected to the helmet. If someone accidentally stepped on the hose, or if the hose got caught on something, the diver's air supply was cut off.

The invention of the **s**elf-**c**ontained **u**nderwater **b**reathing **a**pparatus, better known as **scuba,** allowed divers to carry their air supply on their backs. Ocean explorers Jacques Cousteau and Émile Gagnon improved the scuba system by inventing the Aqua-Lung.

The Aqua-Lung allows a diver to breathe air at a regulated pressure using a mouthpiece. The mouthpiece supplies just the right amount of air from a pressurized tank strapped to the diver's back. With an Aqua-Lung, a diver can safely go down 75 m while breathing compressed air and oxygen (**Figure 2**). Today, anyone can scuba dive and explore underwater areas that were once thought to be impossible to explore.

Figure 2

A scuba diver wears a metal tank that is filled with compressed air. A regulator attaches to the tank. To breathe, the diver inhales air from the regulator, which reduces the pressure of the air to match the surrounding water pressure.

Floating and Sinking— Controlling Buoyancy

Imagine swimming or even floating in a bathtub. The ability to float is called **buoyancy** [BOY-uhn-see]. If you hold your face out of the water and breathe normally, you float at the top of the water. This is because air is lighter, or less dense, than water. The air in your lungs makes you buoyant. If you blow out some of the air, you will sink (**Figure 3**).

Figure 3

When your body has less air and your density is greater than the water, you sink in the water.

A boat floats because its density is less than the density of the water. The water exerts an upward buoyant force on the boat. The boat sinks down into the water until it has displaced, or pushed aside, a volume of water that has the same weight as the boat (**Figure 4**). An object has neutral buoyancy if it neither floats nor sinks.

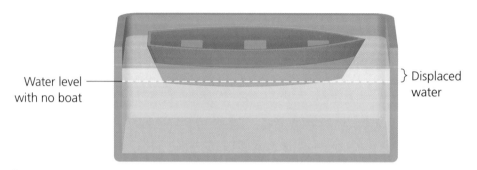

Water level with no boat

} Displaced water

Figure 4
A boat that weighs 1000 kg will sink into the water until it has displaced 1000 kg of water.

TRY THIS: **WHY DOES IT FLOAT?**

Skills Focus: observing, inferring

Fill a large plastic cup about half full with water. Place the cup on a flat surface, and mark the level of the water. Place a piece of wood in the cup, and again mark the water level.

1. What happens to the water level when the wood is in the cup? Why does this happen?

2. What does this tell you about buoyancy?

3. What would happen to the water level if you used different types of wood of equal volume, such as maple and pine? Try this to find out.

Using Technology to Control Buoyancy

If you want to explore underwater, you have to be able to sink and to come back to the surface. To dive under the water, a scuba diver wears a belt that contains weights. They also wear a buoyancy compensator. A buoyancy compensator is a vest that can hold air. To sink, the diver releases air from the vest. To rise again, the diver blows air from the compressed air tank into the vest, filling it with air. This causes the diver to float back up to the surface.

Submarines sink and rise in the water in a similar way. Special tanks, called ballast [BAL-uhst] tanks, in the outer compartment of a submarine can be filled with water or air (**Figure 5**). When the ballast tanks are filled with water, the submarine sinks. When compressed air is pumped back into the ballast tanks, the submarine floats to the surface.

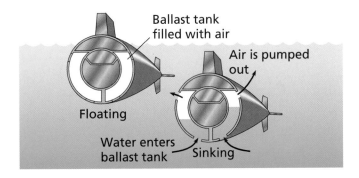

Ballast tank filled with air

Air is pumped out

Floating

Water enters ballast tank

Sinking

Figure 5
Ballast tanks control a submarine's buoyancy.

Surviving Water Pressure in the Ocean Depths

Have you ever felt your ears pop when you dived underwater or swam to the bottom of a pool? This happens because the water above you presses on your body, squeezing the air out of places such as your ears and sinuses. The **water pressure** increases as you go deeper underwater. At just below 10 m, the pressure of the water on your body is twice what the air pressure, or atmospheric pressure, is at the surface. As you go deeper, the pressure continues to increase.

Underwater vehicles must have very strong bodies, or hulls, to withstand the tremendous water pressure at great depths. Submersibles, such as submarines, are pressurized vehicles that have normal air pressure inside. The outside of a submersible is made of titanium [ti-TAY-knee-um]. Titanium is a very strong metal, which is also used to make spacecraft.

Some submersibles, such as the *Alvin* shown in **Figure 6**, can take scientists to a depth of 4500 m. Other submersibles, called remote operated vehicles, or ROVs, do not carry passengers. An ROV is operated with a joystick by an oceanographer on board a large research vessel.

Figure 6
The *Alvin* carries one pilot and two scientists. It has underwater cameras, lights, a television system, and instruments to collect samples from the water.

Canada's Contributions to Ocean Exploration

Canada has a special interest in ocean exploration because of its access to three oceans: the Atlantic, the Pacific, and the Arctic Oceans. Canada has developed technology for exploring its oceans, including a specialized diving suit called a Newt Suit and a submersible called Deep Worker. Canada, along with the United States, is also developing an underwater observatory called the Neptune Project.

The Newt Suit is built to withstand the pressure of deep water (**Figure 7**). Developed by Vancouver diver Phil Nuytten, it looks like something an astronaut would wear underwater. Two electric thrusters are attached to the suit, to move the diver forward. The suit is very heavy out of the water, but nearly weightless underwater. A person wearing the suit can work at 305 m below the surface for up to 8 h.

Figure 7
The Newt Suit is worn by divers who are drilling for oil or gas, or building pipelines for communication companies.

The Deep Worker is a Canadian-designed one-person submersible used to explore the underwater world (**Figure 8**). The Deep Worker is so small that it has been described as an underwater sports car. Explorers are able to go deeper underwater and spend more time underwater than they can with traditional scuba gear.

Figure 8
The Deep Worker can take photos and collect samples from the bottom of the ocean.

Canadian and American scientists are working together on the Neptune Project, which will give scientists a new understanding of deep ocean activity in the Pacific Ocean (**Figure 9**). This new knowledge can be applied to many global issues, such as predicting earthquakes, tracking marine life, understanding climate change, and discovering new energy sources. One of the goals of the Neptune Project is to become a global centre for ocean research.

Figure 9
The Neptune Project will study the diversity of deep-sea life.

⫸ CHECK YOUR UNDERSTANDING

1. What three obstacles make ocean exploration difficult?
2. Why is it important for humans to explore the depths of the oceans?
3. What types of technology have been developed by Canadians to explore the ocean depths?

Figure 1
Giant tube worms can be over 1 m long.

Designing a Sample Collector

One of the most extreme environments on Earth is the ocean floor. In some places, incredibly hot water erupts from cracks, or vents, along the ocean floor. The water can get as hot as 400 °C. Mineral deposits build up around the vents forming stacks. Scientists call these vents black smokers because they look like they are producing black smoke.

Extraordinary creatures, such as giant tube worms (**Figure 1**), live around black smokers. A tube worm has no mouth or gut, but feeds on the bacteria that live inside it. These bacteria can live without oxygen. They do not depend on the Sun for food, since sunlight does not reach the bottom of the ocean. Instead, they break down chemicals from the black smokers to make food for themselves and the tube worms.

Problem

Imagine that you have just joined a team of oceanographers investigating a newly discovered thermal vent ecosystem. You need to design a device for collecting samples from the ocean floor for further study.

Figure 2
ROPOS uses its manipulator arm to collect fluid samples from between rocks on the ocean floor.

Task

Design a device for collecting samples from the area around a thermal vent. Draw a plan for your design and explain how it would work.

Criteria

To be successful, your design must

- show how the device would function under the extreme conditions (such as darkness, and high temperatures and pressures) in and around a thermal vent
- show how a sample would be collected from the ocean floor
- show how the sample would be transported to the surface
- show how the device would be connected to an ROV (**Figure 2**)

Plan and Test

1. Give your device a name.
2. Describe how your device works.
3. Draw a picture of your device, and label all the parts. Describe the function of each part.
4. Decide what samples your device could collect. How would it collect samples in a way that does not disrupt the environment?
5. Describe how the samples would be transported to the surface.
6. Describe any special features you have included to make sure that your device would survive the trip to the bottom of the ocean and back.
7. Explain how your device would be connected to an ROV.

Evaluate

8. Consider the list of criteria. Does your design meet the criteria? Explain.
9. Re-evaluate your design. Are there any changes you would make to your design? Why or why not?

Communicate

10. How will you share your design? You could build a model of your device using simple craft materials or modelling clay. Perhaps you could create a video or computer animation to promote your device.

CHECK YOUR UNDERSTANDING

1. Why are the criteria listed important when designing a new device?
2. What questions should scientists ask before they collect live samples of plants or animals?
3. Explain why it is important to test and re-evaluate your design.

Technology allows us to explore Earth's extreme environments.

Key Idea: Technology allows us to navigate Earth's extreme environments.

Sonar is used to map the ocean floor and to help ships navigate through shallow and rocky waters.

Vocabulary

radar p. 186
satellite p. 186
sonar p. 186

Key Idea: Technology allows us to survive extreme temperatures.

Vocabulary

insulator p. 188

Key Idea: Technology allows us to explore the ocean depths.

Vocabulary

scuba p. 192
buoyancy p. 193
water pressure
   p. 195

Scuba gear and submersibles are two types of technology that were developed to explore the oceans.

Key Idea: Canadians make important contributions to ocean exploration.

The Newt Suit and the Deep Worker are examples of Canadian technology.

Review Key Ideas and Vocabulary

When answering the questions, remember to use the chapter vocabulary.

1. Describe two technologies that have enabled humans to navigate in extreme environments.

2. Draw and label a diagram to show how sonar is used to map the ocean floor.

3. Think about a very cold winter day. Based on your understanding of insulation, what materials would you wear?

4. Describe two Canadian technologies that allow scientists to explore the ocean depths.

5. What do you think is the most important invention for allowing humans to explore underwater? Explain your choice.

Use What You've Learned

6. What are some of the things that humans hope to discover by exploring the ocean? How will these discoveries help society?

7. What is known about the ocean environment? What are some of the unknowns?

www·science·nelson·com GO

8. Make a time line of the various types of technology that are used to explore the oceans.

www·science·nelson·com GO

9. Some Inuit clothing is made from caribou skins. The long, hollow caribou hair is good insulation from the cold. As well, Inuit clothing often has fur on the inside to prevent body heat from escaping into the icy environment. For the same reason, Inuit boots are insulated with fur. Based on what you know about insulation and heat loss, explain why the Inuit use these designs for their clothing and boots.

Think Critically

10. What do you think scientists should consider before they collect living samples from any environment? Explain why.

11. What special features would a deep-ocean collecting device need to have to survive the extreme pressures and temperatures, and to be able to function in the dark?

12. When inventing a new device, it is important for the engineer to know what function the device needs to perform. How was the *form* (the design) of your collection sampler affected by the *function* you wanted it to perform?

Reflect on Your Learning

13. Which of the hands-on activities in this chapter did you find the most interesting? What did the activity help you learn?

Technology allows us to explore the extreme environment of space.

KEY IDEAS

▸ Flight technology allows us to explore the skies.

▸ Rocket technology allows us to travel through Earth's atmosphere and into space.

▸ Canadian scientists make important contributions to space exploration.

▸ Living and working in space requires protection from extreme temperatures, a lack of air pressure, and low gravity.

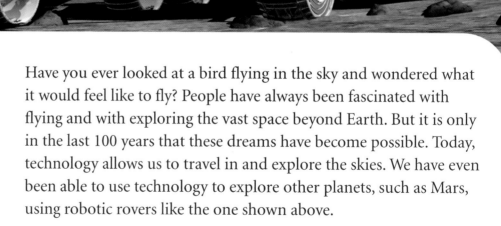

Have you ever looked at a bird flying in the sky and wondered what it would feel like to fly? People have always been fascinated with flying and with exploring the vast space beyond Earth. But it is only in the last 100 years that these dreams have become possible. Today, technology allows us to travel in and explore the skies. We have even been able to use technology to explore other planets, such as Mars, using robotic rovers like the one shown above.

In this chapter, you will learn about the technology of flight and space exploration. You will also discover the important role that Canadian scientists play in exploring space.

Up, Up in the Air

Since ancient times, humans have dreamed of soaring like birds. Hot-air balloons (**Figure 1**), hydrogen balloons, and gliders got people off the ground. Flight depended on the weather, however, and the pilot could not control where the balloon or glider went.

Everything changed in 1903, when Orville and Wilbur Wright achieved the first controlled flight with the *Flyer* (**Figure 2**). A small gasoline motor, attached to the *Flyer's* propellers, moved the plane forward. The Wrights made a total of four flights in the *Flyer*. The longest flight lasted 59 s and covered 260 m. After the final flight, the *Flyer* was overturned by a gust of wind and destroyed.

Figure 1
In 1793, the Montgolfier brothers made a huge paper balloon and filled it with hot air. Since hot air is lighter than cool air, the balloon rose into the sky.

Figure 2
The *Flyer* takes off with Orville Wright at the controls while Wilbur looks on.

After the Wright brothers made their historic flight, different airplane designs made air travel safer and faster. Today, airplanes are made of lightweight metals. They have a streamlined shape and retractable landing gear to reduce air resistance. Millions of people travel the world in high-speed jets. Hot-air balloons and gliders are now used only for recreation.

TRY THIS: HOW DO THE CANS MOVE?

Skills Focus: observing, inferring

Place six straws parallel to each other on a table as shown in **Figure 3**. Space the straws about 2 cm apart. Place two empty aluminum cans on top of the straws. The cans should be about 4 to 6 cm apart. Kneel down so that you are 20 cm from the cans and at eye level with the centre of the cans. Predict what will happen if you blow air in the space between the two cans, using another straw.

Empty cans

Drinking straws

Blow air between cans

Figure 3

 Do not share straws. Each student should use her or his own straw.

1. What happened to the cans? How did they move?
2. Is this what you predicted would happen? Why or why not?
3. Explain why the cans moved the way they did.

Wing Shape and Flight

How does the shape of an airplane's wings help the airplane fly? Imagine that the curve of the aluminum can in the Try This activity is the shape of the top of a wing. Air rushes over the top curve of the wing faster than it moves over the flat bottom surface of the wing. This creates low pressure over the wing, compared with the high pressure under the wing. The high pressure under the wing pushes the wing up and forces the plane upward (**Figure 4**). This is called **lift.** Lift allows people to explore Earth's atmosphere. However, more than lift is needed to get beyond Earth's atmosphere.

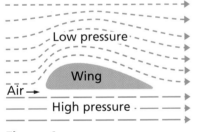

Low pressure

Wing

Air →

High pressure

Figure 4
The movement of air over the wings of an airplane produces lift.

▷ **LEARNING TIP**

After reading "Wing Shape and Flight" on this page, describe to a classmate how wing shape helps an airplane fly. If you have trouble, reread the paragraph and look carefully at **Figure 4** to see how lift is produced.

⫸ CHECK YOUR UNDERSTANDING

1. "The Wright brothers' invention of the airplane changed the world." Do you agree or disagree with this statement. Explain.
2. What do you think is happening to the air pressure between the two aluminum cans when you blow air between them?
3. Why are airplane wings curved?
4. What is lift?

Designing a Super Flyer

Scientists and inventors are problem solvers. The Wright brothers were very good problem solvers. To figure out how to fly, they studied flight, built and rebuilt models, or prototypes, of flying machines, tested and retested their ideas, and redesigned their aircraft. The Wright brothers based their 1903 *Flyer* on their 1902 glider (**Figure 1**).

Figure 1
The Wright 1902 glider was the first controllable aircraft that flew.

In this investigation, you will design, build, test, and redesign a Super Flyer paper airplane.

Problem

How can observations from nature help you design a better paper airplane?

Task

Think about how different animals fly through the air. Then think about how different things in nature, such as seeds, move through the air. How are these things designed to fly? Compare the examples in **Figure 2** with the airplanes in **Figure 3**. How can you use your observations of flight in nature to design an improved paper airplane?

Figure 2
Types of flight in nature

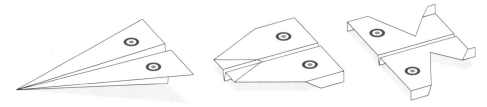

Figure 3
Different designs of paper airplanes

Criteria

To be successful, your final model must

- fly straight for 2 m
- fly at a height of at least 2 m
- be made using the materials you decide upon as a class (e.g., paper, glue, and thin wood)
- incorporate changes based on observations you made when testing and retesting your designs

Plan and Test

1. Work with a partner and design a paper airplane, based on one of the paper airplanes in **Figure 3** or your own ideas. Draw your design.

2. Explain how your airplane will fly. Also explain why you chose to design your airplane the way you did.

3. Prepare a set of instructions for how to build your paper airplane. Use your instructions to build it.

4. Decide how you will test the performance of your paper airplane. For example, consider the following questions:
   - How many times will your repeat your flight tests?
   - What do you plan to observe?
   - What do you plan to measure?

5. Design a table to record your measurements and other observations.

6. Test your paper airplane.

Evaluate

7. How did your airplane do? Based on its performance, design a new airplane that you think will fly farther and straighter. What parts of your design have changed? Why did you change these parts? Modify your diagram and your instructions.

8. Your airplane will participate in a class Paper Airplane Challenge to determine which airplane flies the highest, which airplane flies the farthest, and which airplane flies the straightest.

9. How is your model like a real airplane? How is it different?

Communicate

10. Prepare a labelled diagram, describing how to create your final paper airplane. Include instructions for building your airplane. Give your airplane a catchy name.

> ⯈ *CHECK YOUR* UNDERSTANDING   ⊗
>
> **1.** What is a prototype? Why is it used?
>
> **2.** What did you learn in your flight tests that helped you design your final airplane?
>
> **3.** Why was it important to repeat your tests?

11.3 Rocketing into Space with Technology

The invention and development of the airplane meant that people could explore the skies. Travelling beyond Earth's atmosphere, however, was still not possible. Space is a near vacuum, which means that it contains few oxygen or nitrogen molecules. Since airplane engines need oxygen in the air to burn fuel, airplanes could not be used to explore space.

To explore space, a vehicle that could travel in a vacuum was needed. The vehicle also had to carry its own fuel and source of oxygen. As well, it had to be able to travel fast enough to escape Earth's force of gravity, which pulls things down. The development of rockets, like the one shown in **Figure 1**, and other spacecraft made space exploration possible.

Figure 1
In 1926, Dr. Robert Goddard launched the world's first liquid-fuelled rocket from his backyard.

Before sending people into space, scientists sent unpiloted spacecraft, or probes, into space to gather information and to see what dangers lay ahead. The Russians developed the R-7 rocket to launch the first satellite, *Sputnik I*, into space in 1957. One month later, they launched the first space traveller, a dog named Laika. Then, in 1961, they launched the first human space traveller, Yuri Gagarin. Space technology continued to improve and, in 1969, two American astronauts, Neil Armstrong and Edwin "Buzz" Aldrin, became the first humans to land on the Moon.

Eventually, engineers invented a reusable spacecraft—the space shuttle shown in **Figure 2**. The space shuttle is designed to go into space and then re-enter Earth's atmosphere and land safely. The shuttle commander has only one chance to land successfully. There is only enough fuel to get home and not enough to try the landing again. When the space shuttle touches down, a parachute opens to create drag, or resistance, which helps it stop.

Figure 2
The shuttle is a piloted spacecraft that is launched into space using a rocket.

Payloads and Rockets

Anything that is launched into space, such as a satellite or space shuttle, is called a payload. A launcher carries the payload into space. The main part of the launcher is a rocket.

TRY THIS: *BLAST OFF!*

Skills Focus: creating models, observing, inferring

Blow up a long and skinny balloon, and hold the end closed. Tape a straw along one side of the balloon in a straight line. Feed a piece of fishing line through the straw. Get two assistants to stretch the line across the room and hold the ends tight. Launch the rocket balloon by letting go of the balloon (**Figure 3**).

1. What happens to the balloon? Why does this happen?
2. Compare the rocket balloon with a rocket powered by chemical explosions.

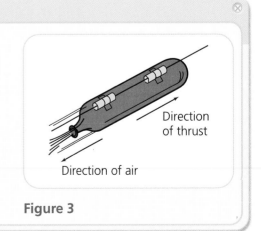

Direction of thrust

Direction of air

Figure 3

A rocket has an opening at one end, like the balloon in the Try This activity. The force of the air escaping from the open end of the balloon moved the balloon forward. In a rocket, the engine mixes fuel with oxygen and produces exhaust gases. The quick release of these gases downward creates an upward force, or **thrust,** on the rocket (**Figure 4**).

Figure 4
The force of the exhaust gases shooting out in one direction causes the thrust of the rocket in the other direction, launching the payload into space.

▷ **LEARNING TIP**

As you read, ask yourself questions to check your understanding: What did I just read? What did it mean? Try to put the information in your own words.

Satellites in Space

Did you know that when you make an overseas phone call, use a cell phone, check the weather forecast, watch TV, or use the Internet, you are using a satellite?

A satellite is an object in space that travels in an orbit around another object. The Moon is a natural satellite of Earth. Human-made satellites are artificial satellites.

Satellites do more than just travel around Earth. Communication satellites receive and transmit television program signals and telephone signals so that you can see and hear about events as they happen, anywhere in the world. For many years, Canada has been an international leader in satellite communication. When the Telesat satellite Anik A1 was launched in 1972, Canada became the first country with its own commercial, domestic communication satellite in orbit. In July 2004, Telesat launched the world's largest commercial communication satellite, Anik F2 (**Figure 5**).

Figure 5
Anik F2 was launched in July 2004. Canada's Telesat communication satellites have all been named Anik, which in the Inuit language means "little brother."

Figure 6
RADARSAT orbits Earth at an altitude of 796 km. It circles Earth 14 times a day. Each orbit takes just over 100 min to complete.

There are many other types of satellites, including navigation, weather, and space research satellites. Some countries even have spy satellites that are used to keep an eye on other countries. Satellites are also used to take pictures of Earth. For example, RADARSAT is a satellite that the Canadian Space Agency developed and operates. RADARSAT provides images of Earth so that scientists can monitor flood damage, soil humidity, forests, and crop conditions, and locate oil spills on the oceans. Images are also used to find surface features that are associated with resources such as oil, water, and minerals. **Figure 6** shows an image of a flooded area of Manitoba, taken using RADARSAT.

Canadian Robots in Space

Canada has made important contributions to space exploration. The Shuttle Remote Manipulator System, commonly known as the **Canadarm,** is a robotic manipulator arm. It was developed by scientists at the Canadian Space Agency and first used on a space shuttle in 1981 (**Figure 7**). An astronaut from inside the space shuttle controls the arm. Over the years, the Canadarm has been changed to adapt to new technology. The Canadarm has been used to send satellites into their proper orbit and retrieve broken satellites for repair, to support space walks by space construction workers, and even to knock ice off the shuttle's wastewater vents.

Figure 7
The Canadarm is 15 m long.

One important job of the Canadarm was the 1993 repair of the Hubble Space Telescope (**Figure 8**). After the Hubble Space Telescope was launched, scientists realized that the photos it took were fuzzy. Using the Canadarm, Hubble was repaired and placed back in orbit to observe the universe.

Figure 8
The Hubble Space Telescope takes computerized pictures of space and sends them to astronomers on Earth.

The Canadarm was also used to install the second robotic arm, the Canadarm 2, on the International Space Station in 2001. While helping to install the new robotic arm, Canadian astronaut Chris Hadfield became the first Canadian to walk in space. Hadfield also manipulated the Canadarm from inside the space shuttle to take a piece of equipment from the Canadarm 2 into the space station—a handshake between two Canadian robotic arms!

⫸ CHECK YOUR UNDERSTANDING ⊗

1. How is thrust created during a rocket launch?
2. Compare an artificial satellite with the Moon. How are they the same? How are they different?
3. What is RADARSAT? What is it used for?
4. Canada has made an important contribution to space exploration through the development of satellites. List two or more things that satellites do for us.
5. What important contribution have Canadian scientists made in robotic space technology? What is this technology used for?

Awesome SCIENCE

Robots can explore where people can't!

NOMAD, THE DESERT EXPLORER

Deserts can be hot, difficult areas to explore, but not for Nomad! Nomad is a four-wheel-drive roving robot, created by scientists at National Aeronautics and Space Administration (NASA). During the 45 days it spent exploring the rugged Atacama Desert in Chile, Nomad travelled 214 km over rough territory—19 km totally on its own. Nomad even picked up an undiscovered rock from the Jurassic Period.

DANTE II, THE VOLCANO EXPLORER

Dante II is a robot that was built by NASA to explore Mount Spurr, an active volcano in Alaska. Dante II was actually the second robot built to explore volcanoes. The first robot, Dante I, was sent into Mount Erebus in Antarctica. It had only covered a few metres before its cable broke and it fell into the inferno below. Unfortunately, Dante II was also damaged as it explored Mount Spurr. Dante II now tours the United States promoting robotic research.

ROPOS AND THE UNDERWATER VOLCANOES

Have you ever wondered what an underwater volcano looks like? ROPOS, which stands for **R**emotely **O**perated **P**latform for **O**cean **S**cience, is a robotic submersible that was developed in Canada. It is used to study submarine volcano systems. It is also used to collect samples and take scientific readings.

ROPOS is placed inside a cage and lowered 5000 m. It is attached to a tether cable and is moved out of the cage to complete the dive. ROPOS uses different tools, such as giant steel jaws, cutters, hooks, and drills, to break off samples from the ocean floor. It uses suction samplers, strainers, and scoops to collect the samples. It has cameras and video and computer recorders to log its underwater explorations. Four people work from the surface ship to direct ROPOS. A scientist directs the dive, a pilot operates the machine, a sample collector operates the tools for collecting the samples, and an event recorder keeps a log of what ROPOS does.

SPIRIT AND OPPORTUNITY ON MARS

In January 2004, two robot geologists—Spirit and Opportunity—landed on the surface of Mars and began to explore their surroundings. Guided from Earth by a team of engineers (in charge of navigating the area) and a team of scientists (in charge of learning about the Martian environment), the rovers travelled across Mars. Spirit took microscopic pictures of an intriguing rock known as Pot-of-Gold. Opportunity moved down into a crater known as Endurance Crater, making observations and using a rock abrasion tool to dig into rocks. Scientists used the computer systems on the two rovers and the computer systems in the laboratory on Earth to analyze soil and rock samples for signs of past life and water on Mars.

SKILLS MENU

- ○ Questioning ● Observing
- ○ Predicting ● Measuring
- ○ Hypothesizing ○ Classifying
- ○ Designing Experiments ● Inferring
- ● Controlling Variables ● Interpreting Data
- ○ Creating Models ● Communicating

Rocket Blasters

Young scientists, here is your chance to discover how to achieve the greatest thrust for a canister rocket. You will work as part of a three-member team. Your teacher will assign your team one of the Questions to investigate.

To make sure that your tests are fair and accurate, pay close attention to the procedure and change only one variable at a time. Do at least three trials, and conduct your tests the same way in each trial. Using the results for each variable, you will collaborate with your team to design a new procedure to create a rocket blaster!

Questions

a) Can I create the greatest thrust with my canister rocket using cold water or warm water?

b) Can I create the greatest thrust with my canister rocket using half of an antacid tablet or a whole tablet?

Materials

- apron
- safety goggles
- 2 film canisters, with lids that fit inside the rims
- paper
- tape
- 6 antacid tablets
- beaker of cold water and beaker of warm water (for Question (a))
- beaker of water at room temperature (for Question (b))
- shoebox lid
- tape measure
- stopwatch

apron

goggles

film canisters and lids

paper

tape

antacid tablets

beaker of water at room temperature

beakers of cold and warm water

shoebox lid

stopwatch

tape measure

> Conduct this experiment outside or in a large indoor space. Wear goggles, and make sure that the canister is placed upside down (inverted) during the tests. Everyone should stand well back from the canisters because the rockets can travel up to 7 m.

Procedure

1 Make a paper rocket (**Figure 1**). Wrap a piece of paper around the film canister and tape it in place. The lid of the canister should face down. Remember to leave room at the bottom of the canister so you can put the lid back on. Cut a piece of paper into a cone and tape it to the top of the rocket. Draw and cut out four fins for your rocket, and tape them in place.

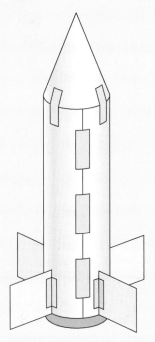

Figure 1

2 Put on your apron and safety goggles. Get six antacid tablets from your teacher. If you are investigating Question (a), you will need a beaker of warm water and a beaker of cold water. If you are investigating Question (b), you will need a beaker of water at room temperature. You will also need to break two antacid tablets in half.

3 Build a launch pad using the shoebox lid. Prop up the lid on an angle. The canister rockets will blast sideways instead of straight up when launched at this angle.

Question (a)

4 Make a table like the one below.

Observations for Question (a)

Water	Time to blast off (s)			Distance travelled (cm)		
	Trial 1	Trial 2	Trial 3	Trial 1	Trial 2	Trial 3
hot						
cold						

5 Fill one canister one-third full of warm water. Drop in one tablet, and quickly put the lid on the canister. Place your rocket on the launch pad as shown in **Figure 2**, and stand back. Measure the time the rocket takes to blast off. Record this time in your table. After the rocket lands, measure the distance it travelled. Record your measurements in your table.

Figure 2

6 Repeat step 5, but this time use cold water. Record your measurements in your table.

7 Repeat steps 5 and 6 two more times, for a total of three trials.

Question (b)

8 Make a table like the one below.

Observations for Question (b)

Amount of tablet	Time to blast off (s)			Distance travelled (cm)		
	Trial 1	Trial 2	Trial 3	Trial 1	Trial 2	Trial 3
one tablet						
half of a tablet						

9 Fill both canisters one-third full of water. Put one tablet in one of the canisters. Quickly put the lid on the canister. Place your rocket on the launch pad, and stand back. Measure the time the rocket takes to blast off. After your rocket has launched, measure the distance your rocket travelled. Record your measurements in your table.

10 Repeat step 9 using half a tablet. Record your measurements in your table.

11 Repeat steps 9 and 10 two more times, for a total of three trials.

▷ **LEARNING TIP**

For help preparing your graph, read "Graphing Data" in the Skills Handbook.

Analyze and Evaluate

1. Find the average time and distance for the three tests.

2. Create a graph to illustrate the results of your test.

3. What conditions created the greatest thrust for your canister rocket?

Apply and Extend

4. As a whole class, discuss the results for Questions (a) and (b). Which conditions do you think will produce the greatest thrust for the canister rockets?

5. What did you do in your investigation to ensure that your results were as accurate as possible?

6. Can you think of other ways to measure the range of your rocket? Are there other variables that you would like to test?

⫸ CHECK YOUR UNDERSTANDING ⊗

1. Why is it important to follow the procedure the same way in each trial? Why is it important to repeat a test?

2. What variable did you change in your investigation?

3. What made your test a fair test?

The International Space Station  11.5

Space stations are important for space exploration. Before humans can live on places like the Moon or Mars, scientists need to understand how space affects the human body. Scientists research the effects of space in space stations that orbit Earth. The International Space Station, or ISS, is the biggest technological project in space (**Figure 1**). The ISS is so important that space agencies from 16 countries around the world are involved in the project. When it is finished, it will be over 100 m long and will orbit about 320 km above Earth.

Figure 1
The International Space Station (ISS) orbits high above Earth.

The ISS is a laboratory in space. Science experiments on the ISS can last for months. The ISS has a microgravity environment, in which the effects of gravity are very small. Everything appears to be almost weightless. Scientists on the ISS are studying the effects of microgravity on animals and plants.

Scientists on the ISS are also studying the effects of space on human bones. As well, they are studying Earth's climate, learning about the solar system, and developing new technologies that may be used for further space exploration. Canadian scientists are conducting experiments that may lead to the development of new medicines and ways to keep astronauts healthy when they are living and working in space.

Once the ISS is complete, scientists hope that it can also be used as a refuelling station for longer missions. If you look up at night, you may see the ISS passing overhead.

⫸ CHECK YOUR UNDERSTANDING

1. Describe some uses of the ISS.
2. Why is it important to research the effects of space on humans and on the growth of plants?

11.6 Living and Working in Space

▷ **LEARNING TIP**

Set a purpose for your reading of this section. First, read the headings and look at the illustrations in the section. Then make a list of questions that you have about how astronauts live and work in space.

In the International Space Station, the phrase "floating off to sleep" has a totally different meaning. Since the ISS has so little gravity, astronauts must attach their sleeping bags to a wall or seat to prevent them from floating around the cabin as they sleep (**Figure 1**). What other things do you think would be difficult to do in space?

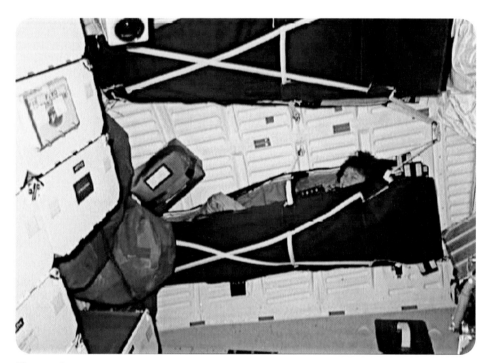

Figure 1
An astronaut zips up for a cozy sleep in space.

Living and working in space is very difficult for astronauts. There are many challenges, such as lack of air in space, extremely low temperatures, and low gravity. Even growing food and getting water are difficult in space.

Breathing in Space

There is no air in space. On the ISS, however, astronauts can breathe easily. This is because of the **life-support systems** on the ISS. The life-support systems provide oxygen for the astronauts to breathe and absorb the carbon dioxide that the astronauts exhale.

Wearing Space Suits for Protection

Space suits are like mini spacecraft, designed to protect astronauts from exposure to space. For example, the normal human body temperature is about 37 °C. When you are hot, your body sweats to cool off. When you are cold, your body warms up by shivering. The temperature of an object in space can drop to −157 °C in the darkness and soar to 121 °C in the sunlight. Astronauts need to wear space suits to protect themselves from these extreme temperatures.

On board the ISS, astronauts wear a T-shirt and shorts as they work and exercise. During a launch or re-entry, however, they wear a partially pressurized suit called a Launch and Entry Space Suit, or LES. An LES can provide enough air pressure to return to Earth during an emergency landing, when cabin pressure in the space shuttle may decrease (**Figure 2**).

Figure 2
The Launch and Entry Space Suit (LES) has batteries for power and radio devices for communication. The LES is also insulated and has an emergency oxygen system, a parachute harness and parachute pack, 2 L of drinking water, and floatation devices.

Figure 3
An astronaut wearing an EMU carries the life-support system in a backpack.

To work in space, an astronaut wears an Extravehicular Mobility Unit, or EMU (**Figure 3**). An EMU has many layers to protect the astronaut from the vacuum of space, extreme temperatures, and the Sun's harmful radiation. A gold visor on the helmet protects the astronaut's eyes from the blinding sunlight. Oxygen tanks provide oxygen for about seven or eight hours. Gloves have heaters to prevent the astronaut's hands from freezing when the astronaut is working in space at night in the cold temperatures. The EMU also contains a small thermostat that can be adjusted if the astronaut becomes too warm while working in space during the day. Tubes coil through special underwear to keep the astronaut at a comfortable temperature. Not surprisingly, the EMU is very heavy. It is made with ball bearings at the joints, which allow the astronaut to bend and twist.

The Manned Maneuvering Unit, or MMU, is a nitrogen-propelled backpack that latches to the EMU and allows the astronaut to move when outside of the spacecraft. An astronaut wearing an MMU can move forward, backward, turn, and even do flips in space.

Falling in Space

If you drop an apple on Earth, it falls to the ground. What happens if an astronaut on the ISS drops an apple? Even though the apple may look like it is floating, just like Chris Hadfield appears to be floating in **Figure 4**, it is actually falling. In fact, the apple, the astronaut, and the ISS are all falling together around Earth. Since they are all falling at the same rate, the apple and the astronaut appear to be weightless inside the ISS. The microgravity condition in the ISS makes them appear to float.

Figure 4
It looks like Canadian astronaut Chris Hadfield is floating. He is actually falling.

Skills Focus: creating models, observing, inferring

The reason that objects feel weightless in space is because they are falling toward Earth's surface. You can make a model to show weightlessness (**Figure 5**). Start by cutting a rubber band. Put one end so that it hangs down into a plastic bottle, while the other end lies across the mouth of the bottle. Screw on the lid to keep the rubber band in place. Note your observations about the rubber band. Unscrew the lid, and put some modelling clay on one end of the rubber band. Put the end with the modelling clay in the bottle. The other end should drape across the mouth of the bottle. Screw the lid on the bottle. Notice the effect of the modelling clay on the rubber band. Hold the bottle about 1 m in the air, and drop it.

1. What do you observe?
2. How do your observations relate to what happens on the ISS?

Figure 5

Because of microgravity, there is a lack of force against the muscles in an astronaut's body. When astronauts are living on the ISS, their muscles become smaller and their bones lose calcium and become weak. The spine and other joints spread apart. This causes the astronauts to stretch up to 5 cm taller. To combat the effects of microgravity on their bodies, astronauts exercise daily on special exercise machines (**Figure 6**). They even race from one end of the ISS to the other to keep in shape!

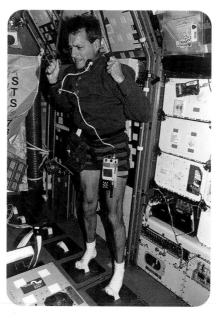

Figure 6
This astronaut is exercising while on the space shuttle.

Food, Water, and Waste in Space

Survival in space depends on having enough food and water, and on finding ways to dispose of waste. Astronauts get three meals a day. The food must be nutritious and easy to eat in a floating environment (**Figure 7**). The astronauts have more than 100 foods to choose from, including fruits, nuts, peanut butter, chicken, beef, seafood, and candy. Drinks include coffee, tea, orange juice, fruit punch, and lemonade. Drinking straws have clamps that stop the liquid from coming out after the astronaut stops sucking. Some foods are dehydrated [dee-HI-dray-ted], so the astronauts just need to add water. Most foods are precooked, so the astronauts just reheat them in an oven. There are no refrigerators on the ISS.

Figure 7
Astronauts have to be careful that pieces of food do not escape and get into the sensitive instruments.

Did you know that when you exhale, or breathe out, your breath contains water? On the ISS, all water—including water from the astronauts' breath—is recycled and purified so that it can be used again. There is a lot of water on the ISS in containers that were transported from Earth. Since bringing water from Earth is expensive, scientists have developed technology to collect humidity from the air. In the future, every drop of water, from waste water to water used for hygiene, may be recycled and purified on the ISS. On Earth, when we **recycle,** we reuse something instead of discarding it. On the ISS, it is very important to find ways to recycle all materials and to reduce waste as much as possible.

Designing materials so they can be reused is important. Most garbage is packed up and returned to Earth, but sometimes it is deposited into space, where it burns up. Staying clean on the space station is a must. Micro-organisms (bacteria) grow easily in a closed system. Astronauts do not get very dirty in space. To stay clean, they use a moist cloth to wipe themselves clean.

What about personal wastes? To use a toilet on the ISS, astronauts first strap themselves in so they will not float away (**Figure 8**). They then sit on a rubber ring to create a seal around the toilet as their solid waste is vacuumed into a waste receptacle. There is a hose to collect urine. All liquids, including urine, are processed to remove pure water, which can be reused.

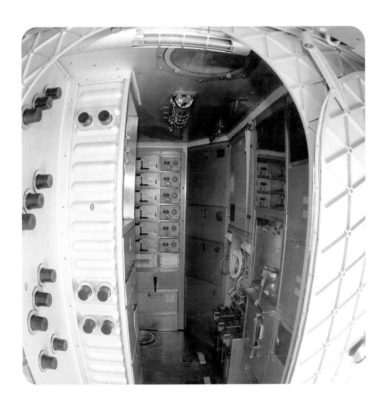

Figure 8
The toilet on the ISS is quite a bit different from the toilets on Earth.

▐▶ CHECK YOUR UNDERSTANDING

1. How are astronauts able to breathe on the ISS?
2. What are some functions of a space suit?
3. Water is a valuable resource on Earth and on the ISS. Compare how you use water to how an astronaut uses water.
4. Why is it important for astronauts to exercise in space?
5. Compare the needs of an astronaut on a space flight for a few days with the needs of an astronaut living in space for a few months.

11.6 Living and Working in Space

11 Chapter Review

Technology allows us to explore the extreme environment of space.

Key Idea: Flight technology allows us to explore the skies.

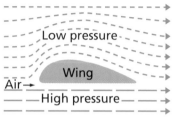

Curved wings give an airplane lift so it can fly.

Vocabulary

lift p. 204

Key Idea: Rocket technology allows us to travel through Earth's atmosphere and into space.

The space shuttle

Vocabulary

thrust p. 210

Key Idea: Canadian scientists make important contributions to space exploration.

The Canadarm and RADARSAT are Canadian contributions to space exploration.

Vocabulary

Canadarm p. 212

Key Idea: Living and working in space requires protection from extreme temperatures, a lack of air pressure, and low gravity.

The International Space Station is a laboratory in space.

Vocabulary

life-support systems p. 220

recycle p. 224

Review Key Ideas and Vocabulary

When answering the questions, remember to use the chapter vocabulary.

1. How did an understanding of lift help people develop the technology of flight?

2. Give an example of a flying machine, such as a helicopter, that humans have invented. What things or creatures from nature may have inspired the design?

3. How does thrust cause a rocket to launch into space?

4. What are the three main obstacles to human survival in space? For each obstacle, discuss a technology that has made it possible for humans to survive in space.

5. Describe two important Canadian contributions to space exploration.

Use What You've Learned

6. What happens if one member of a scientific team does not complete his or her job properly? For example, in Investigation 11.4, there were three members on each team. Write a paragraph about the importance of completing a job properly. Infer how your experiences in this investigation could relate to space exploration.

7. Research the countries that are involved in the International Space Station.

www.science.nelson.com

8. Research Canada's contributions to the ISS.

www.science.nelson.com

9. Research a Canadian astronaut. When did this astronaut go into space? What did he or she do on the mission?

www.science.nelson.com

10. On October 5, 2004, *SpaceShipOne* became the first privately built manned rocket ship to fly into space (**Figure 1**). In the future, people may be able to fly into space on board rocket ships like *SpaceShipOne*. What do you think about the possibility of riding in a rocket ship? What could this mean for the future of space exploration?

Figure 1
SpaceShipOne and its launch ship

Think Critically

11. What lessons about recycling on Earth can be implemented to reduce waste in space?

12. What do you think would be the most difficult part of living in space? Why?

Reflect on Your Learning

13. Which activity best helped you understand the concepts presented in this chapter? How do you learn best?

14. In this chapter, you have learned about the exploration of space, the most extreme environment. What aspect of space exploration interested you the most? What questions do you have about future space explorations?

CHAPTER 12

Exploring extreme environments has both benefits and costs.

KEY IDEAS

▸ The exploration of extreme environments produces spinoffs and other benefits.

▸ The exploration of extreme environments has drawbacks and costs.

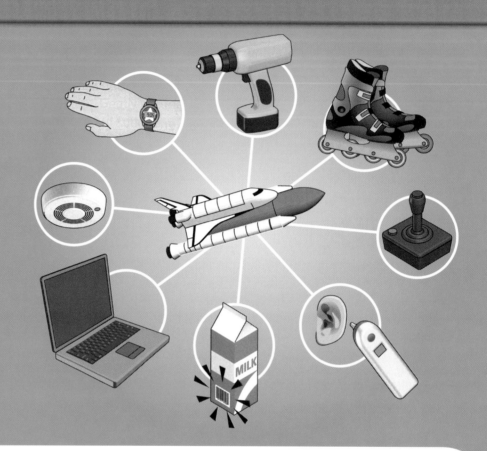

Did you know that the joystick you use to play video games is modelled on the controls that astronauts use to practise shuttle landings? Exploration technology is all around you. Many inventions created for exploring extreme environments find their way into everyday life.

Exploration provides knowledge about the world you live in. New species of plants and animals discovered deep in the oceans may lead to new medicines. Satellites in space are used to predict the weather on Earth. Information sent from sensors on the ocean floor may help predict earthquakes.

But there are costs to exploring extreme environments, too. Deciding whether the benefits of exploration outweigh the costs is an important issue, not just for scientists, but for everyone.

What do the bar codes you see on many products have in common with space technology? National Aeronautics and Space Agency (NASA) developed bar code technology to keep track of millions of spacecraft parts. Today, grocery stores and department stores use bar codes to keep track of their products (**Figure 1**). Bar code technology is an example of a **spinoff** —an everyday use of a technology that was first developed for exploration.

Figure 1
Scanning bar codes is just one example of space technology.

Space exploration, in particular, has given us many spinoffs. Smoke detectors, like the ones used in your home, were originally developed to detect deadly gases on *Skylab*, the first space station. Cordless tools were originally developed for astronauts to use on the Moon to collect rock samples. Even portable laptop computers were first used on space shuttle missions. Other examples of spinoffs are shown in **Figure 2**.

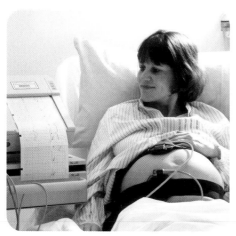 

Figure 2
A type of fetal heart monitor (left) and industrial robots that put cars together (right) are two spinoffs from space technology.

Spinoffs from space and ocean exploration are listed in **Table 1**.

Table 1 Spinoffs from Space and Ocean Exploration

Exploration technology	Examples
microelectronics	digital watches, computers, heart pacemakers, calculators, cordless tools
new materials	waterproof materials, flame-resistant materials, nonstick coating
ceramic materials	dental braces
plastics	safety helmets, in-line skates
space food	freeze-dried foods
robotics	building cars, mining, oil exploration
medicine	motion sickness patches, scanning equipment, fetal heart monitor, heart pump, kidney dialysis, insulin pumps, temperature pill, surgical probe

TRY THIS: RESEARCH A SPINOFF

Skills Focus: questioning, inferring

Look at the spinoff technologies listed in **Table 1**. Choose one technology and research how it was developed. For example, you could research the development of in-line skates or safety helmets (**Figure 3**). Make a time line for the technology. How is this technology used in daily life?

Figure 3
Helmets and in-line skates are spinoffs of exploration technology.

⫸ CHECK YOUR UNDERSTANDING

1. What is a spinoff? Why are spinoffs important?
2. What exploration spinoff do you think has the most impact on your life? Explain your choice.

Exploration and Your Health

The technology developed for space exploration has produced many medical spinoffs. For example, heart pacemakers, laser surgery, and medical imaging systems are three important spinoffs from space exploration.

We also benefit from medical experiments done in space. In space, substances mix together more easily, and crystals grow differently because of the microgravity environment. These conditions allow scientists to develop new medicines. Space medicines have been used to treat people on Earth who have diabetes, burns, and blood diseases. Research is also being done on producing medicines that are made with "space-grown" crystals.

Three Canadian astronauts—Dr. Roberta Bondar (**Figure 1**), Dr. Robert Thirsk, and Dr. Dave Williams (**Figure 2**)—are medical doctors who specialize in space medicine. They study the impact of working and living in space. As well, they experiment to find ways to apply what they have learned to help people on Earth. For example, astronauts are given medication to stop them from feeling dizzy during landing. This medication is now being used to treat heart patients on Earth.

Figure 1
Roberta Bondar flew on the space shuttle *Discovery* in 1992. The crew looked at how microgravity affects shrimp eggs, lentil seedlings, and bacteria. Dr. Bondar also investigated how humans adapt to weightlessness.

Figure 2
Dave Williams flew on the space shuttle *Columbia* in 1998. During the 17-day mission, the crew studied the effects of weightlessness on the nervous system. They also looked at how the inner ear, cardiovascular system, and muscles cope without gravity.

⫸ CHECK YOUR UNDERSTANDING

1. Give two reasons why scientists do medical experiments in space.
2. Why is it important to study the effects that living in space has on astronauts?

12.3 The Drawbacks of Exploration

The discovery of new species, spinoffs, knowledge about the world and the universe, and new energy sources are just some of the benefits of exploring extreme environments. However, there are also drawbacks to explorations.

Space exploration, for example, is costly and dangerous. Since space exploration began, 18 astronauts have been killed on missions. The most recent space disaster occurred on February 2003, when the space shuttle *Columbia* (**Figure 1**) disintegrated upon re-entry to Earth's atmosphere. All seven of the astronauts on board were killed. Even training for space missions can be deadly. Ten astronauts have been killed in training accidents on the ground.

What orbits Earth, is found on the surfaces of the Moon, Venus, and Mars, and is greatly feared by space explorers? Space junk! Space junk includes broken satellites, discarded pieces of rockets, and even nuts and bolts from spacecraft. All human-made objects that remain in orbit and serve no useful purpose are called space junk. Space scientists estimate that there are millions of pieces of space junk floating around Earth (**Figure 2**).

Figure 1
The space shuttle *Columbia*

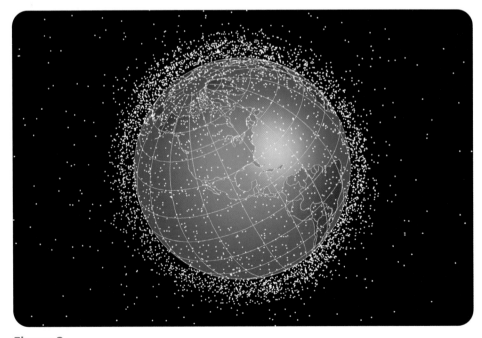

Figure 2
Space junk orbits Earth and poses a hazard to the International Space Station and to spacecraft.

Space junk comes in different sizes and shapes. Some pieces are as large as trucks. Others are smaller than a flake of paint. There's even a glove that floated away from the crew of *Gemini 4* during the first spacewalk and a camera lost by an astronaut during the *Gemini 10* mission. All this space junk zooms around Earth at speeds of up to 36 000 km/h.

Space junk poses a danger to working satellites and spacecraft because it travels at great speeds. Even small pieces can damage a spacecraft in a collision (**Figure 3**).

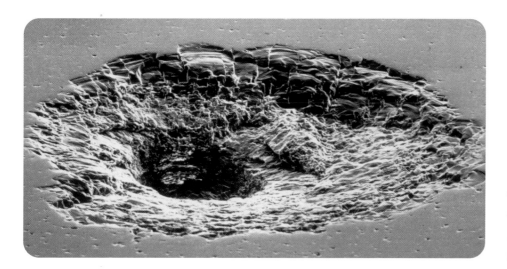

Figure 3
This tiny crater in the window of the space shuttle *Challenger* may have been caused by a flake of paint.

Ground stations track large pieces of space junk so that collisions with working satellites, spacecraft, and the International Space Station can be avoided. Different countries are also working on plans to stop the creation of space junk and to clean up what is already there. Perhaps a future job may be space junk collector!

Another drawback to exploration is cost. Billions of dollars are spent designing and testing vehicles for space and ocean exploration. This money could, perhaps, be better spent elsewhere. Or could it? In the next section, you will look at whether exploration is worth the cost.

▮▶ CHECK YOUR UNDERSTANDING

1. What are some of the drawbacks of space exploration?
2. What is space junk? Where does space junk come from?
3. What drawbacks do you think are associated with exploring volcanoes or oceans?

Is Exploration Worth the Cost?

Exploration benefits us in many ways, from the spinoff technologies that we use in daily life to the discovery of new resources. However, the exploration of extreme environments is costly, and scientists don't always know whether an exploration will prove worthwhile.

The Issue

Some people believe that exploration, especially space exploration, costs too much. They would prefer to see the money used to deal with problems that are closer to home, such as social problems and pollution (**Figure 1**).

Figure 1
Some people believe that the money spent on exploring space should be spent on cleaning up polluted rivers and lakes.

Background to the Issue

Designing and developing the technology needed to explore extreme environments is very costly. For example, Canada's space program receives about $300 million from the federal government. Even though this is only a small amount of Canada's national budget, it represents a lot of money. People who support space and ocean exploration believe that the spinoffs and benefits outweigh the cost. What do you think? Is the exploration of extreme environments worth the cost?

Identify Perspectives

To decide whether the exploration of extreme environments is worth the cost, you must look at the issue from different perspectives. This means that you have to look at both the positive aspects and the negative aspects of exploration.

Working in groups of three, choose one of the extreme environments that you have learned about in this unit: polar regions, deserts, oceans, volcanoes, or space.

Use the Internet and the library to research your topic. Start by looking for information on recent expeditions to study the extreme environment that you have chosen. Make sure that you look at both the benefits and drawbacks of your topic. Use questions, such as the following, to help identify the information you need:

- What were the goals of the expeditions?
- What benefits could the expeditions provide?
- What technology did the exploration require?
- What were the problems, dangers, and costs of the exploration?

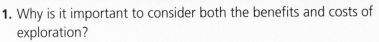
www.science.nelson.com

How do you feel about about your topic after assessing the information that you found? Can you conclude whether this type of exploration is worth the cost?

Communicate Your Ideas

Share what you have learned with your class. You can make a presentation, design a poster, act out a scene, give a speech about your topic, or use a graphic organizer to present what you have learned.

> ### ▐▶ CHECK YOUR UNDERSTANDING ⊗
>
> 1. Why is it important to consider both the benefits and costs of exploration?
> 2. What difficulties did you encounter in trying to weigh the benefits and the costs of exploration?

LEARNING TIP ◁

When you look at the benefits and drawbacks of something, you are looking at the pros and cons. What other topics have you learned about that were organized into pros and cons?

12 *Chapter Review*

Exploring extreme environments has both benefits and costs.

Key Idea: The exploration of extreme environments produces spinoffs and other benefits.

Vocabulary

spinoff p. 229

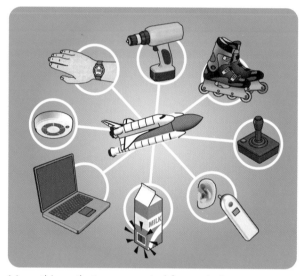

Many things that were created for exploring space are used in everyday life.

Key Idea: The exploration of extreme environments has drawbacks and costs.

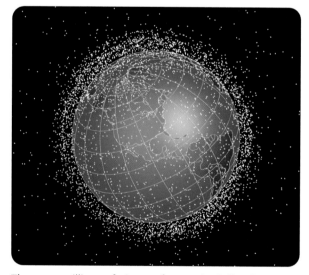

There are millions of pieces of space junk floating around Earth.

Review Key Ideas and Vocabulary

When answering the questions, remember to use the chapter vocabulary.

1. What are some of the benefits of exploring extreme environments?

2. How are spinoffs related to space exploration? Give an example of a spinoff.

3. What are some of the drawbacks of exploring extreme environments?

Use What You've Learned

4. Think about different careers that are related to a spinoff of science exploration. Which career most interests you?

5. Look at **Figure 1**. What space missions are taking place right now? Do an Internet search to find out. Select one mission, and write a short description of its purpose.

www·science·nelson·com

Figure 1
The Cassini-Huygens mission to Saturn began on October 15, 1997. It reached Saturn's rings on June 30, 2004.

6. Name three Canadian astronauts who have conducted health experiments on space shuttle missions. Research one of these astronauts.

www·science·nelson·com

7. New medicines are being created in space. Research and write a paragraph about the work of Canadian astronauts in space health sciences.

www·science·nelson·com

Think Critically

8. Dogs and chimpanzees were launched into space before humans. Many of the experiments done in the International Space Station involve fish, birds, snakes, frogs, rats, jellyfish, and insects. Why are animals sent into space? Does the benefit to humans outweigh the risk to animals? Discuss your answers in a small group.

Reflect on Your Learning

9. What spinoffs of exploration technology most surprised you? Why?

10. It is hard to believe that 50 years ago, video games did not exist and no one imagined working as a video-game programmer. Invent a career that you think will exist as a result of exploration technology when you are an adult.

Making Connections

Design an Exploration Vehicle

Looking Back

In this unit, you learned about extreme environments and the obstacles they present for human survival. You learned that technology allows people to travel to and survive in extreme conditions, like those found in space and in the oceans (**Figure 1**).

For this project, you will work with a partner to design and build a 3-D model of an exploration vehicle to travel to and investigate an extreme environment.

Demonstrate Your Learning

Part 1: Research an extreme environment
Which extreme environment would you like to explore? Choose one of the environments discussed in this chapter: polar regions, deserts, oceans, volcanoes, or space. Research the challenges that this environment would pose for an exploration vehicle. Also research the scientific principles that need to be incorporated into your design to overcome these challenges. For example, a submersible

Figure 1
The *Johnson Sea-Link* is a submersible used to study life in deep water, perform search and recovery tasks, and conduct underwater archeological missions.

would need to have a pressurized hull to withstand the immense water pressure. A vehicle that is travelling over rough terrain would need large wheels (**Figure 2**).

Figure 2
NASA is developing Big Wheels, an inflatable robot rover, for future missions to Mars. Its large wheels will allow it to travel over rocky terrain.

Part 2: Brainstorm the design of an exploration vehicle

Brainstorm different ideas for the design of your exploration vehicle. Consider the following questions:

- What information do you want to collect about the extreme environment you are exploring?

- How will your vehicle gather this information? (For example, how will it take pictures or collect samples?) What devices and accessories will you need to collect scientific data?

- How is your vehicle controlled and guided? (For example, does it need a driver or is it remote controlled?)

- How will your vehicle move?

Draw a diagram of the model you want to build. Discuss the materials you will need to build your model.

Part 3: Build a model of your exploration vehicle

Build a model of your exploration vehicle. Use your diagram as a guide. Make any adjustments to your design as necessary.

Part 4: Communicate

Prepare a report about your exploration vehicle. Describe the environment in which your vehicle would be used. Also describe the function and purpose of your vehicle.

⫸ ASSESSMENT ⊗

MODEL

Check to make sure that your model of an exploration vehicle provides evidence that you are able to

- apply appropriate technology
- design a model that meets the conditions of an extreme environment
- work cooperatively with a partner

REPORT

Check to make sure that your report provides evidence that you are able to

- identify the information needed
- find appropriate sources of information
- identify and describe the conditions in an extreme environment
- use appropriate scientific language
- communicate clearly

SKILLS HANDBOOK

THINKING AS A SCIENTIST

You may not think you're a scientist, but you are! You investigate the world around you, just like scientists do. When you investigate, you are looking for answers. Imagine that you are planning to buy a mountain bike. You want to find out which model is the best buy. First, you write a list of questions. Then you visit stores, check print and Internet sources, and talk to your friends to find the answers. You are conducting an investigation.

Scientists conduct investigations for different purposes:

- *Scientists investigate the natural world in order to describe it.* For example, oceanographers explore the deep ocean to study the habitats of deep-sea creatures and to look for new organisms.

- *Scientists investigate how objects and organisms can be classified.* For example, biologists examine organisms to see if they have one cell or many cells. By classifying organisms on the basis of their similarities and differences, scientists are able to organize their observations and learn about the relationships among living things.

- *Scientists investigate to test their ideas about the natural world.* Scientists ask cause-and-effect questions about what they observe. They propose hypotheses to answer their questions. Then they design experiments to test their hypotheses.

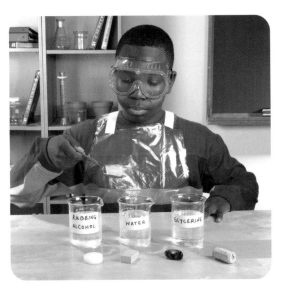

CONDUCTING AN INVESTIGATION

When you conduct an investigation or design an experiment, you will use a variety of skills. Refer to this section when you have questions about how to use any of the following investigation skills and processes.

- Questioning
- Predicting
- Hypothesizing
- Controlling Variables
- Observing
- Measuring
- Classifying
- Inferring
- Interpreting Data
- Communicating
- Creating Models

Questioning

Scientific investigations start with good questions. To write a good question, you must first decide what you want to know.

You must think carefully about what you want to know in order to develop a good question. The question should include the information you want to find out.

Sometimes an investigation starts with a special type of question, called a cause-and-effect question. A cause-and-effect question asks whether something is causing something else. It might start in one of the following ways:
What causes ...?
How does ... affect ...?
What would happen if ...?

When an investigation starts with a cause-and-effect question, it also has a hypothesis. Read "Hypothesizing" on page 244 to find out more about hypotheses.

PRACTICE

Think of some everyday examples of cause and effect, and write statements about them. Here's one example: "When I stay up too late, I'm tired the next day." Then turn your statements into cause-and-effect questions: for example, "What happens if I stay up late?"

Predicting

A prediction states what is likely to happen based on what is already known. Scientists base their predictions on their observations. They look for patterns in the data they gather to help them see what might happen next or in a similar situation. This is how meteorologists come up with weather forecasts.

Remember that predictions are not guesses. They are based on solid evidence and careful observations. You must be able to give reasons for your predictions. You must also be able to test them by doing experiments.

Hypothesizing

To test your questions and predictions scientifically, you need to conduct an investigation. Use a question or prediction to create a cause-and-effect statement that can be tested. This kind of statement is called a **hypothesis**.

An easy way to make sure that your hypothesis is a cause-and-effect statement is to use the form "If … then …" (**Figure 1**). For example, "If the number of times a balloon is rubbed against hair (the cause) is increased, then the length of time it sticks to a wall (the effect) increases."

When you conduct an investigation, you do not always prove that your hypothesis is correct. Sometimes you prove that your hypothesis is incorrect. An investigation that proves your hypothesis to be incorrect is not a bad investigation or a waste of time. It has contributed to your scientific knowledge. You can re-evaluate your hypothesis and design a new experiment.

Figure 1

This student is conducting an investigation to test this hypothesis: if the number of times the balloon is rubbed against hair increases, then the length of time it will stick to the wall will increase.

PRACTICE

Write hypotheses for questions or predictions about rubbing a balloon on your hair and sticking it to a wall. Start with the questions above, and then write your own questions. For example, if your question is "Does the balloon stick better if you rub it more times?", then your hypothesis might be "If the number of times you rub the balloon on your hair is increased, then the length of time it sticks to the wall is increased."

Controlling Variables

When you are planning an investigation, you need to make sure that your results will be reliable by conducting a fair test. To make sure that an investigation is a fair test, scientists identify all the variables that might affect their results. Then they make sure that they change only one variable at a time. This way they know that their results are caused by the variable they changed and not by any other variables (**Figure 2**).

- The variable that is changed in an investigation is called the **independent variable**.

- The variable that is affected by a change is called the **dependent variable**. This is the variable you measure to see how it was affected by the independent variable.

- All the other conditions that remain unchanged in an experiment, so that you know they did not have any effect on the outcome, are called **controlled variables**.

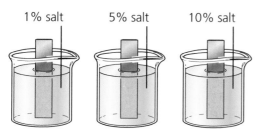

Figure 2

This investigation was designed to find out if the amount of salt in a solution has an effect on the rusting of metal.

- The amount of salt in each solution is the independent variable.
- The amount of rust on the pieces of metal is the dependent variable.
- The amount of water in each beaker and the amount of time the metal strip stays in the water are two of the controlled variables.

Observing

When you observe something, you use your senses to learn about the world around you. You can also use tools, such as a balance, metre stick, and microscope.

Some observations are measurable. They can be expressed in numbers. Observations of time, temperature, volume, and distance can all be measured. These types of observations are called **quantitative observations**.

Other observations describe qualities that cannot be measured. The smell of a fungus, the shape of a flower petal, or the texture of soil are all examples of qualities that cannot be put in numbers. These types of observations are called **qualitative observations**. Qualitative observations also include colour, taste, clarity, and state of matter.

The colour and shape of this box are qualitative observations. The measurements of its height, depth, and width are quantitative observations.

Measuring

Measuring is an important part of observation. When you measure an object, you can describe it precisely and keep track of any changes. To learn about using measuring tools, turn to "Measurement and Measuring Tools" on page 259.

Measuring accurately requires care.

Classifying

You classify things when you sort them into groups based on their similarities and differences. When you sort clothes, sporting equipment, or books, you are using a classification system. To be helpful to other people, a classification system must make sense to them. If, for example, your local supermarket sorted all the products in alphabetical order, so that soap, soup, and soy sauce were all on the same shelf, no one would be able to find anything!

Classification is an important skill in science. Scientists try to group objects, organisms, and events in order to understand the nature of life (**Figure 3**).

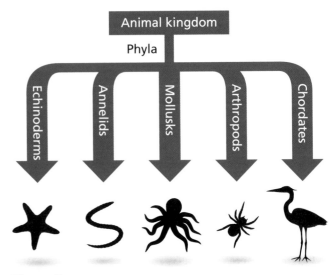

Figure 3
To help classify animals, scientists divide the animal kingdom into five smaller groups called *phyla* (singular *phylum*).

PRACTICE

Gather photos of 15 to 20 different types of insects, seashells, or flowers. Try to include as much variety as possible. How are all your samples alike? How are they different? How could you classify them?

Inferring

An inference is a possible explanation of something you observe. It is an educated guess based on your experience, knowledge, and observations. You can test your inferences by doing experiments.

It is important to remember that an inference is only an educated guess. There is always some uncertainty. For example, if you hear a dog barking but do not see the dog, you may infer that it is your neighbour's dog. It may, however, be some other dog that sounds the same. An observation, on the other hand, is based on what you discover with your senses and measuring tools. If you say that you heard a dog barking, you are making an observation.

PRACTICE

Decide whether each of these statements is an observation or an inference.

a) You see a bottle filled with clear liquid. You conclude that the liquid is water.

b) You notice that your head is stuffed up and you feel hot. You decide that you must have a cold.

c) You tell a friend that three new houses are being built in your neighbourhood.

d) You see a wasp crawling on the ground instead of flying. You conclude that it must be sick.

e) You notice that you are thirsty after playing sports.

Interpreting Data

When you interpret data from an investigation, you make sense of it. You examine and compare the measurements you have made. You look for patterns and relationships that will help you explain your results and give you new information about the question you are investigating. Once you have interpreted your data, you can tell whether your predictions or hypotheses are correct. You may even come up with a new hypothesis that can be tested in a new experiment.

Often, making tables or graphs of your data will help you see patterns and relationships more easily (**Figure 4**). Turn to "Communicating in Science" on page 271 to learn more about creating data tables and graphing your results.

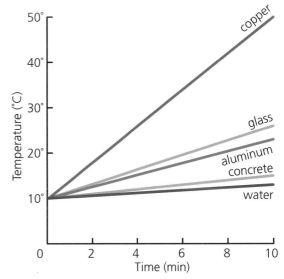

Figure 4
This graph shows data from an investigation about the heating rates of different materials. What patterns and relationships can you see from this data?

Communicating

Scientists learn from one another by sharing their observations and conclusions. They present their data in charts, tables, or graphs and in written reports. In this student text, each investigation or activity tells you how to prepare and present your results. To learn more about communicating in a written report, turn to "Writing a Lab Report" on page 274.

Creating Models

Have you ever seen a model of the solar system? Many teachers use a small model of the solar system when teaching about space because it shows how the nine planets orbit the Sun.

A scientific model is an idea, illustration, or object that represents something in the natural world (**Figure 5**). Models allow you to examine and investigate things that are very large, very small, very complicated, very dangerous, or hidden from view. They also allow you to investigate processes that happen too slowly to be observed directly. You can model, in a few minutes, processes that take months or even millions of years to occur.

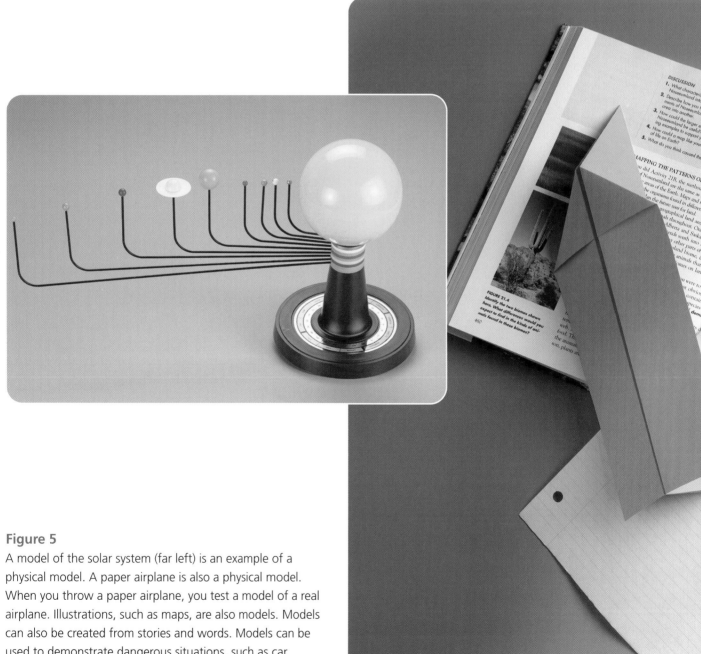

Figure 5
A model of the solar system (far left) is an example of a physical model. A paper airplane is also a physical model. When you throw a paper airplane, you test a model of a real airplane. Illustrations, such as maps, are also models. Models can also be created from stories and words. Models can be used to demonstrate dangerous situations, such as car crashes (far right). Can you think of any other models?

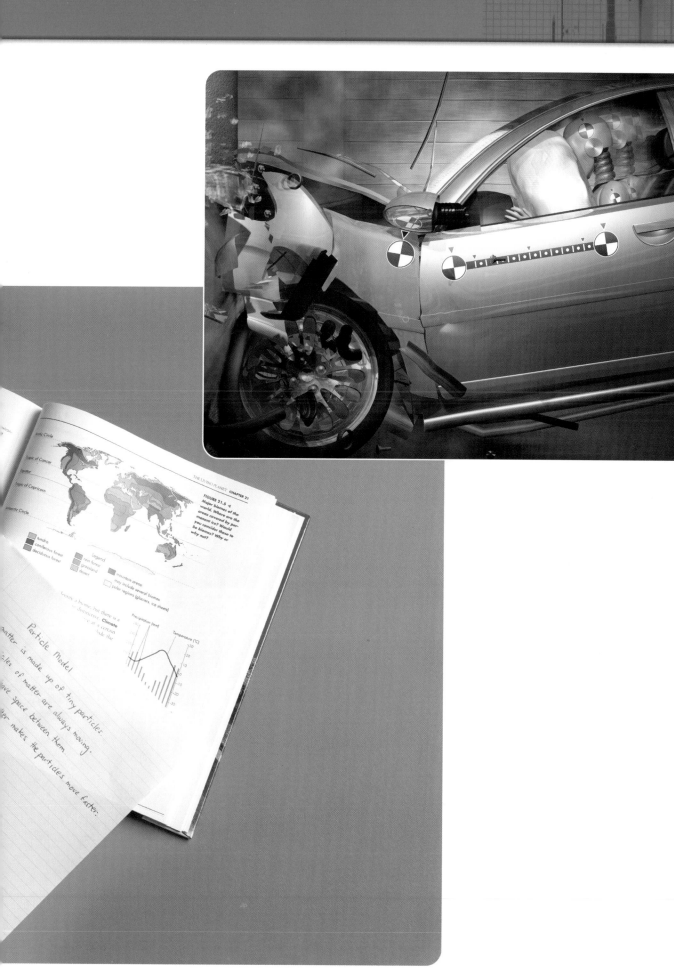

SOLVING A PROBLEM

Refer to this section when you are doing a "Solve a Problem" activity.

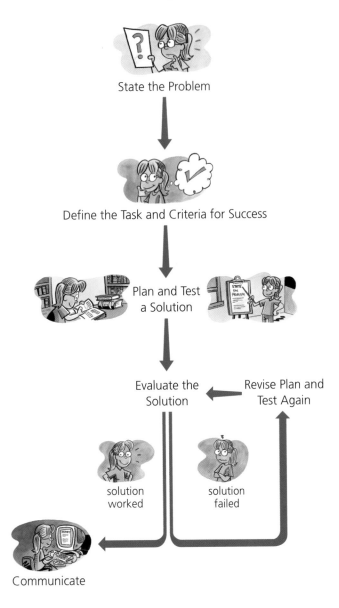

State the Problem

Define the Task and Criteria for Success

Plan and Test a Solution

Evaluate the Solution

Revise Plan and Test Again

solution worked

solution failed

Communicate

State the Problem

The first step in solving a problem is to state what the problem is. Imagine, for example, that you are part of a group that is investigating how to reduce the risk of people getting the West Nile Virus. People can become very sick from this virus.

When you are trying to understand a problem, ask yourself these questions:

- What is the problem? How can I state it as a problem?
- What do I already know about the problem?
- What do I need to know to solve the problem?

Define the Task and the Criteria for Success

Once you understand the problem, you can define the task. The task is what you need to do to find a solution. For the West Nile Virus problem, you may need to find a way to reduce the number of mosquitoes in your community because they could be carrying the West Nile Virus.

Before you start to consider possible solutions, you need to know what you want your solution to achieve. One of the criteria for success is fewer mosquitoes. Not every solution that would help you achieve success will be acceptable, however. For example, some chemical solutions may kill other, valuable insects or may be poisonous to birds and pets. The solution should not be worse than the problem it is meant to solve. As well, there are limits on your choices. These limits may include the cost of the solution, the availability of materials, and safety.

Use the following questions to help you define your task and your criteria for success:

- What do I want my solution to achieve?
- What criteria should my solution meet?
- What are the limits on my solution?

Plan and Test a Solution

The planning stage is when you look at possible solutions and decide which solution is most likely to work. This stage usually starts with brainstorming possible solutions. When you are looking for solutions, let your imagination go. Keep a record of your ideas. Include sketches, word webs, and other graphic organizers to help you.

As you examine the possible solutions, you may find new questions that need to be researched. You may want to do library and Internet research, interview experts, and talk to people in your community about the problem.

Choose one solution to try. For the West Nile Virus problem, you may decide to inspect your community for wet areas where mosquitoes breed, and try to eliminate as many of these wet areas as possible. You have discovered, through your research, that this solution is highly effective for reducing mosquito populations. It also has the advantage of not involving chemicals and costing very little.

Now make a list of the materials and equipment you will need. Develop your plan on paper so that other people can examine it and add suggestions. Make your plan as thorough as possible so that you have a blueprint for how you are going to carry out your solution. Show your plan to your teacher for approval.

Once your teacher has approved your plan, you need to test it. Testing allows you to see how well your plan works and to decide whether it meets your criteria for success. Testing also tells you what you might need to do to improve your solution.

Evaluate the Solution

The evaluating stage is when you consider how well your solution worked. Use these questions to help you evaluate your solution:

- What worked well? What did not work well?
- What would I do differently next time?
- What did I learn that I can apply to other problems?

If your solution did not work, go back to your plan and revise it. Then test again.

Communicate

At the end of your problem-solving activity, you should have a recommendation to share with others. To communicate your recommendation, you need to write a report. Think about what information you should include in your report. For example, you may want to include visuals, such as diagrams and tables, to help others understand your results and recommendation.

DESIGNING YOUR OWN EXPERIMENT

Refer to this section when you are designing your own experiment.

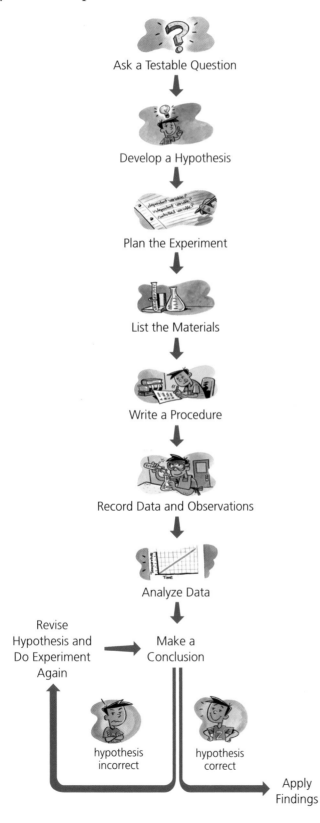

Ask a Testable Question

Develop a Hypothesis

Plan the Experiment

List the Materials

Write a Procedure

Record Data and Observations

Analyze Data

Revise Hypothesis and Do Experiment Again → Make a Conclusion

hypothesis incorrect

hypothesis correct

Apply Findings

After observing the difference between his lunch and Dal's, Simon wondered why his food was not as fresh as Dal's.

Scientists design experiments to test their ideas about the things they observe. They follow the same steps you will follow when you design an experiment.

Ask a Testable Question

The first thing you need is a testable question. A testable question is a question that you can answer by conducting a test. A good, precise question will help you design your experiment. What question do you think Simon, in the picture above, would ask?

A testable question is often a cause-and-effect question. Turn to "Questioning" on page 243 to learn how to formulate a cause-and-effect question.

Develop a Hypothesis

Next, use your past experiences and observations to formulate a hypothesis. Your hypothesis should provide an answer to your question and briefly explain why you think the answer is correct. It should be testable through an experiment. What do you think Simon's hypothesis would be? Turn to "Hypothesizing" on page 244 to learn how to formulate a hypothesis.

Plan the Experiment

Now you need to plan how you will conduct your experiment. Remember that your experiment must be a fair test. Also remember that you must only change one independent variable at a time. You need to know what your dependent variable will be and what variables you will control. What do you think Simon's independent variable would be? What do you think his dependent variable would be? What variables would he need to control? Turn to "Controlling Variables" on page 244 to learn about fair tests and variables.

List the Materials

Make a list of all the materials you will need to conduct your experiment. Your list must include specific quantities and sizes, where needed. As well, you should draw a diagram to show how you will set up the equipment. What materials would Simon need to complete his experiment?

Write a Procedure

The procedure is a step-by-step description of how you will perform your experiment. It must be clear enough for someone else to follow exactly. It must explain how you will deal with each of the variables in your experiment. As well, it must include any safety precautions. Your teacher must approve your procedure and list of materials. What steps and safety precautions should Simon include?

Record Data and Observations

You need to make careful observations, so that you can be sure about the effects of the independent variable. Record your observations, both qualitative and quantitative, in a data table, tally chart, or graph. How would Simon record his observations?

Turn to "Observing" on page 245 to read about qualitative and quantitative observations. Turn to "Creating Data Tables" on page 271 to read about creating data tables.

Analyze Data

If your experiment is a fair test, you can use your observations to determine the effects of the independent variable. You can analyze your observations to find out how the independent and dependent variables are related. Scientists often conduct the same test several times to make sure that their observations are accurate.

Make a Conclusion

When you have analyzed your observations, you can use the results to answer your question and determine if your hypothesis was correct. You can feel confident about your conclusion if your experiment was a fair test and there was little room for error. If you proved that your hypothesis was incorrect, you can revise your hypothesis and perform the experiment again.

Apply Findings

The results of scientific experiments add to our knowledge about the world. For example, the results may be applied to develop new technologies and medicines, which help to improve our lives. How do you think Simon could use what he discovered?

PRACTICE

You are a tennis player. You observe that your tennis ball bounces differently when the court is wet. Design a fair test to investigate your observation. Use the headings in this section.

EXPLORING AN ISSUE

Use this section when you are doing an "Explore an Issue" activity.

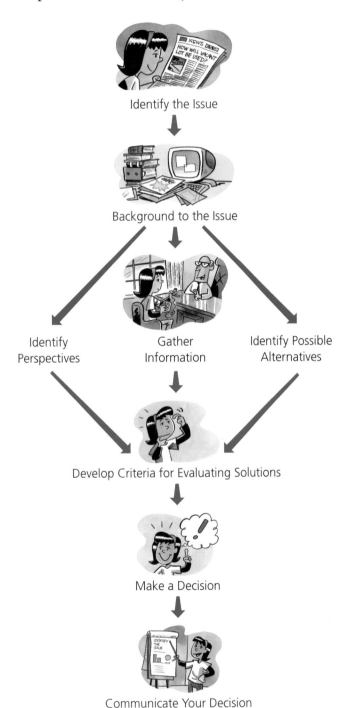

Identify the Issue

Background to the Issue

Identify Perspectives

Gather Information

Identify Possible Alternatives

Develop Criteria for Evaluating Solutions

Make a Decision

Communicate Your Decision

You make decisions everyday that can affect yourself, others, and the environment. What might seem like the right decision to you can look quite different to someone else. For example, you might think that bicycles are the only means of transportation that people need. But many other people rely on their cars and could not replace them with bicycles.

People hold different viewpoints about a lot of issues that affect other people and the environment. An issue is a situation in which several points of view need to be considered in order to make a decision. Often what different people think is the best decision is based on what they think is important or on what they value. Often, it is difficult to come to a decision that everyone agrees with.

When a decision has an impact on many people or on the environment, it is important to explore the issue carefully. This means thinking about all the possible solutions and trying to understand all the different points of view—not just your own point of view. It also means researching and investigating your ideas, and talking to and listening to others.

Identify the Issue

The first step in exploring an issue is to identify what the issue is. An issue has more than one solution, and there are different points of view about which solution is the best. Try stating the issue as a question: "What should ...?"

Background to the Issue

The background to the issue is all the information that needs to be gathered and considered before a decision can be made.

- *Identify perspectives.* There are always different points of view on an issue. That's what makes it an issue. For example, suppose that your municipal council is trying to decide how to use some vacant land next to your school. You and other students have asked the council to zone the land as a nature park. Another group is proposing that the land be used to build a seniors' home because there is a shortage of this kind of housing. Some school administrators would like to use the land to build a track for runners and sporting events.

- *Gather information.* The decision you reach must be based on a good understanding of the issue. You must be in a position to choose the most appropriate solution. To do this, you need to gather factual information that represents all the different points of view. Watch out for biased information, presenting only one side of the issue. Develop good questions and a plan for your research. Your research may include talking to people, reading about the issue, and doing Internet research. For the land-use issue, you may also want to visit the site to make observations.

- *Identify possible alternatives.* After identifying points of view and gathering information, you can now generate a list of possible solutions. You might, for example, come up with the following solutions for the land-use issue:
  - Turn the land into a nature park for the community and the school.
  - Use the land as a playing field and track for the community and the school.

- Create a combination park and playing field.
- Use the land to build a seniors' home, with a "nature" garden.

Develop Criteria for Evaluating Solutions

Develop criteria to evaluate each possible solution. For example, should the solution be the one that has the most community support? Should it be the one that protects the environment? You need to decide which criteria you will use to evaluate the solutions so that you can decide which solution is the best.

Make a Decision

This is the stage where everyone gets a chance to share his or her ideas and the information he or she gathered about the issue. Then the group needs to evaluate all the possible solutions and decide on one solution based on the list of criteria.

Communicate Your Decision

Choose a method to communicate your decision. For example, you could choose one of the following methods:

- Write a report.
- Give an oral presentation.
- Design a poster.
- Prepare a slide show.
- Create a video.
- Organize a panel presentation.
- Write a newspaper article.
- Hold a formal debate.

WORKING AS A SCIENTIST

GETTING OFF TO A SAFE START

Science activities and investigations can be a lot of fun. You have the chance to work with new equipment and substances. These can be dangerous, however, so you have to pay attention! You also have to know and follow special rules. Here are the most important rules to remember.

1 **Follow your teacher's directions.**

- Listen to your teacher's directions, and follow them carefully.

- Ask your teacher for directions if you are not sure what to do.

- Never change anything, or start an activity on your own, without your teacher's approval.

- Get your teacher's approval before you start an experiment that you have designed yourself.

2 **Act responsibly.**

- Pay attention to your own safety and the safety of others.

- Tell your teacher immediately if you see a safety hazard, such as broken glass or a spill. Also tell your teacher if you see another student doing something that you think is dangerous.

- Tell your teacher about any allergies or medical problems you have, or about anything else your teacher should know.

- Do not wear contact lenses while doing experiments.

- Read all written instructions carefully before you start an activity.

- Clean up and put away any equipment after you are finished.

3 **Be science-ready.**

- Come prepared with your student text, notebook, pencil, worksheets, and anything else you need for an activity or investigation.

- Keep yourself and your work area tidy and clean.

- Wash your hands carefully with soap and water at the end of each activity or investigation.

- Never eat, drink, or chew gum in the science classroom.

- Wear safety goggles or other safety equipment when instructed by your teacher.

- Keep your clothing and hair out of the way. Roll up your sleeves, tuck in loose clothing, and tie back loose hair. Remove any loose jewellery.

SAFE SCIENCE

Follow these instructions to use chemicals and equipment safely in the science classroom.

HEAT, FIRE, AND ELECTRICITY

- Never heat anything without your teacher's permission.
- Always wear safety goggles when you are working with fire.
- Keep yourself, and anything else that can burn, away from heat or flames.
- Never reach across a flame.
- Before you heat a test tube or another container, point it away from yourself and others. Liquid inside can splash or boil over when heated.
- Never heat a liquid in a closed container.
- Use tongs or heat-resistant gloves to pick up a hot object.
- Test an object that has been heated before you touch it. Slowly bring the back of your hand toward the object to make sure that it is not hot.
- Know where the fire extinguisher and fire blanket are kept in your classroom.
- Never touch an electrical appliance or outlet with wet hands.
- Keep water away from electrical equipment.

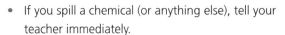

CHEMICALS

- If you spill a chemical (or anything else), tell your teacher immediately.
- Never taste, smell, touch, or mix chemicals without your teacher's permission.
- Never put your nose directly over a chemical to smell it. Gently wave your hand over the chemical until you can smell the fumes.
- Keep the lids on chemicals you are not using tightly closed.
- Wash your hands well with soap after handling chemicals.
- Never pour anything into a sink without your teacher's permission.
- If any part of your body comes in contact with a chemical, wash the area immediately and thoroughly with water. If your eyes are affected, do not touch them but wash them immediately and continuously with cool water for at least 15 min. Inform your teacher.

HANDLE WITH CARE

GLASS AND SHARP OBJECTS

- Handle glassware, knives, and other sharp instruments with extra care.
- If you break glassware or cut yourself, tell your teacher immediately.
- Never work with cracked or chipped glassware. Give it to your teacher.
- Use knives and other cutting instruments carefully. Never point a knife or sharp object at another person.
- When cutting, make sure that you cut away from yourself and others.

LIVING THINGS

- Treat all living things with care and respect.
- Never treat an animal in a way that would cause it pain or injury.
- Touch animals only when necessary. Follow your teacher's directions.
- Always wash your hands with soap after working with animals or touching their cages or containers.

Caution Symbols

The activities and investigations in *B.C. Science Probe 6* are safe to perform, but accidents can happen. This is why potential safety hazards are identified with caution symbols and red type (**Figure 1**). Make sure you read the cautions carefully and understand what they mean. Check with your teacher if you are unsure.

 Wash your hands with soap and water after each time you work with the plants.

Figure 1
Potential safety hazards are identified with caution symbols and red type.

Safety Symbols

The following safety symbols are used throughout Canada to identify products that can be hazardous (**Figure 2**). Make sure that you know what each symbol means. Always use extra care when you see any of these symbols in your classroom or anywhere else.

PRACTICE

In a group, create a safety poster for your classroom. For example, you could create a map of the route your class should follow when a fire alarm sounds, a map of where safety materials (such as a fire extinguisher and a first-aid kit) are located in your classroom, information about the safe use of a specific tool, or a list of safety rules.

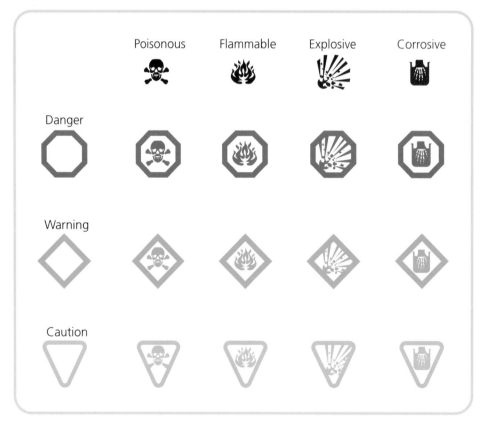

Figure 2
Hazardous Household Product Symbols (HHPS) appear on many products that are used in the home. Different shapes show the level of danger.

MEASUREMENT AND MEASURING TOOLS

Refer to this section when you need help with taking measurements.

Measuring is an important part of doing science. Measurements allow you to give exact information when you are describing something.

These are the most commonly used measurements:

- Length
- Mass
- Volume
- Temperature

The science community and most countries in the world, including Canada, use the SI system. The SI system is commonly called the metric system.

The metric system is based on multiples of 10. Larger and smaller units are created by multiplying or dividing the value of the base units by multiples of 10. For example, the prefix *kilo-* means "multiplied by 1000." Therefore, one kilometre is equal to one thousand metres. The prefix *milli-* means "divided by 1000," so one millimetre is equal to $\frac{1}{1000}$ of a metre. Some common SI prefixes are listed in **Table 1**.

Table 1 Common SI Prefixes

Prefix	Symbol	Factor by which unit is multiplied	Example
kilo	k	1000	1 km = 1000 m
hecto	h	100	1 hm = 100 m
deca	da	10	1 dam = 10 m
		1	
deci	d	0.1	1 dm = 0.1 m
centi	c	0.01	1 cm = 0.01 m
milli	m	0.001	1 mm = 0.001 m

Measuring Length

Length is the distance between two points. Four units can be used to measure length: kilometres (km), metres (m), centimetres (cm), and millimetres (mm).

10 mm = 1 cm	100 cm = 1 m
1000 mm = 1 m	1000 m = 1 km

You measure length when you want to find out how long something is. You also measure length when you want to know how deep, how tall, how far, or how wide something is. The metre is the basic unit of length (**Figure 3**).

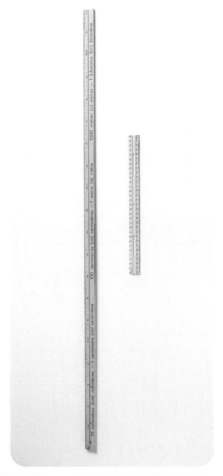

Figure 3
Metre sticks can be used to measure long lengths, up to 100 cm. Metric rulers are used to measure shorter lengths in millimetres and centimetres, up to 30 cm.

PRACTICE

Which unit—millimetres, centimetres, metres, or kilometres—would you use to measure each quantity?

a) the width of a scar or mole on your body

b) the length that your toenails grow in one month

c) your height

d) the length that your hair grows in one month

e) the distance between your home and Calgary

f) the distance between two planets

Tips for Measuring Length

- Always start measuring from the zero mark on a ruler, not from the edge of the ruler.

- Look directly at the lines on the ruler. If you try to read the ruler at an angle, you will get an incorrect measurement.

- To measure something that is not in a straight line, use a piece of string (**Figure 4**). Cut or mark the string. Then use a ruler to measure the length of the string. You could also use a tape measure made from fabric.

Figure 4

Measuring Volume

Volume is the amount of space that something takes up. The volume of a solid is usually measured in cubic metres (m^3) or cubic centimetres (cm^3). The volume of a liquid is usually measured in litres (L) or millilitres (mL).

1000 mL = 1 L	1 L = 1000 cm^3
1 cm^3 = 1 mL	1000 L = 1 m^3

The volume of a rectangular solid is calculated by measuring the length, width, and height of the solid and then by using the formula

$$\text{volume} = \text{length} \times \text{width} \times \text{height}$$

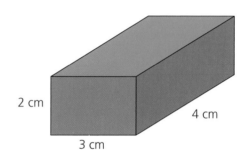

Volume is also used to measure the amount of liquid in a container. Scientists use special containers, such as beakers and graduated cylinders, to get precise measurements of volume.

You can also use liquid to help measure the volume of irregularly shaped solids, such as rocks. To measure the volume of an irregularly shaped solid, choose a container (such as a graduated cylinder) that the irregular solid will fit inside. Pour water into the empty container until it is about half full. Record the volume of water in the container, and then carefully add the solid. Make sure that the solid is completely submerged in the water. Record the volume of the water plus the solid. Calculate the volume of the solid using the following formula:

$$\text{volume of solid} = (\text{volume of water} + \text{solid}) - \text{volume of water}$$

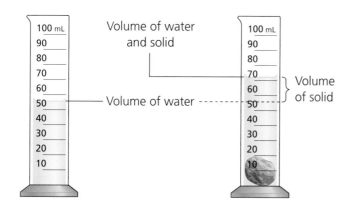

Volume of water and solid

Volume of water

Volume of solid

PRACTICE

What volume of liquids do you drink in an average day? Use the illustrations of volume measurements to help you answer this question.

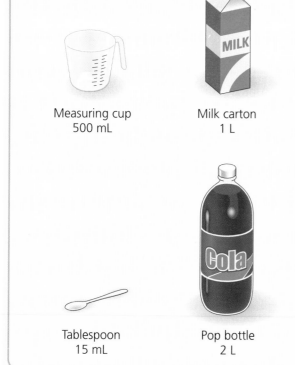

Measuring cup
500 mL

Milk carton
1 L

Tablespoon
15 mL

Pop bottle
2 L

Tips for Measuring Volume

- Use a beaker that is big enough to hold twice as much liquid as you need. You want a lot of space so that you can get an accurate reading.

- To measure liquid in a graduated cylinder (or a beaker or a measuring cup), make sure that your eyes are at the same level as the top of the liquid. You will see that the surface of the liquid curves downward. This downward curve is called the **meniscus.** You need to measure the volume from the bottom of the meniscus (**Figure 5**).

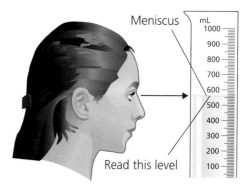

Meniscus

Read this level

Figure 5
Reading the measurement of a liquid correctly

- Use a graduated cylinder to get the most accurate measurement of volume.

Measuring Mass

Mass is the amount of matter in an object. In everyday life, weight is often confused with mass. For example, you probably state your weight in kilograms. In fact, what you are really stating is your mass. The units that are used to measure mass are grams (g), milligrams (mg), kilograms (kg), and metric tonnes (t).

1000 g = 1 kg   1000 kg = 1 t
1000 mg = 1 g

Scientists use balances to measure mass. Two types of balances are the triple-beam balance (**Figure 6**) and the platform, or equal-arm, balance (**Figure 7**).

Measuring Temperature

Temperature is the degree of hotness or coldness of an object. In science, temperature is measured in degrees Celsius.

0 °C = freezing point of water
20 °C = room temperature
37.6 °C = normal body temperature
65 °C = water hot to touch
100 °C = boiling point of water

Figure 6
A triple-beam balance: Place the object you are measuring on the pan. Adjust the weights on each beam (starting with the largest) until the pointer on the right side is level with the zero mark. Then add the values of each beam to find the measurement.

Figure 7
A platform balance: Place the object you are measuring on one pan. Add weights to the other pan until the two pans are level. Then add the values of the weights you added. The total will be equal to the mass of the object you are measuring.

Tips for Measuring Mass
- To measure the mass of a liquid, first measure the mass of a suitable container. Then measure the mass of the liquid in the container. Subtract the mass of the container from the mass of the liquid and the container.

- To measure the mass of a powder or crystals, first determine the mass of a sheet of paper. Then place the sample on the sheet of paper, and measure the mass of both. Subtract the mass of the paper from the mass of the sample and the sheet of paper.

Measuring the temperature of water

Each mark on a Celsius thermometer is equal to one degree Celsius. The glass contains a coloured liquid—usually mercury or alcohol. When you place the thermometer in a substance, the liquid in the thermometer moves to indicate the temperature.

Tips for Measuring Temperature
- Make sure that the coloured liquid has stopped moving before you take your reading.

- Hold the thermometer at eye level to be sure that your reading is accurate.

READING FOR INFORMATION

USING GRAPHIC ORGANIZERS

Diagrams that are used to organize and display ideas visually are called graphic organizers. A graphic organizer can help you see connections and patterns among different ideas. Different graphic organizers are used for different purposes.

- To Show Processes
- To Organize Ideas and Thinking
- To Compare and Contrast
- To Show Properties or Characteristics

To Show Processes

You can use a **flow chart** to show a sequence of steps or a time line.

This diagram shows a feeding pathway or food chain—urchins eat kelp, otters eat urchins, and orcas eat otters.

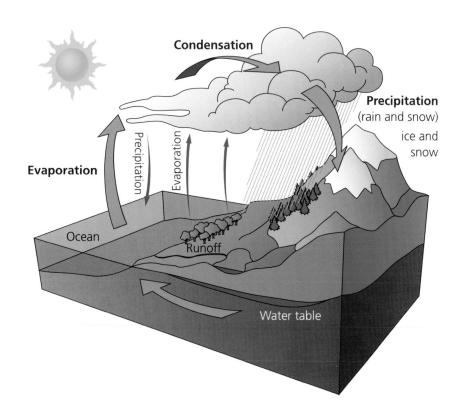

You can use a **cycle map** to show cycles in nature.

This diagram shows the water cycle.

To Organize Ideas and Thinking

A **concept map** is a collection of words or pictures, or both, connected with lines or arrows. You can write on the lines or arrows to explain the connections. You can use a concept map to brainstorm what you already know, to mind map your thinking, or to summarize what you have learned.

This concept map shows who eats whom in an ecosystem.

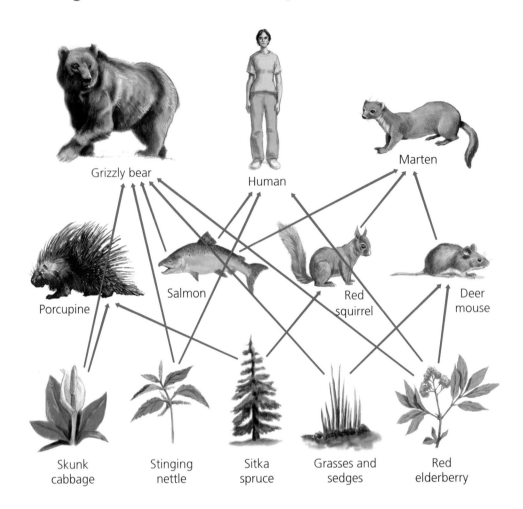

Grizzly bear

Human

Marten

Porcupine

Salmon

Red squirrel

Deer mouse

Skunk cabbage

Stinging nettle

Sitka spruce

Grasses and sedges

Red elderberry

You can use a **tree diagram** to show concepts that can be broken down into smaller categories.

The Animal kingdom can be broken down into smaller groups called phyla.

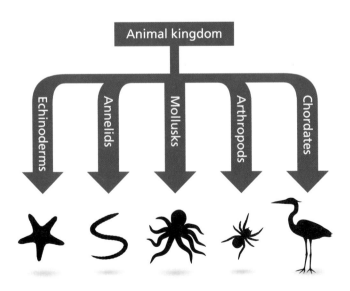

Animal kingdom

Echinoderms

Annelids

Mollusks

Arthropods

Chordates

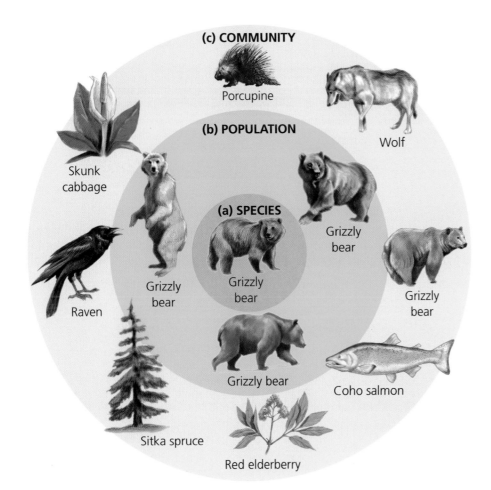

(c) COMMUNITY

Porcupine

Wolf

(b) POPULATION

Skunk cabbage

(a) SPECIES

Grizzly bear

Grizzly bear

Grizzly bear

Grizzly bear

Grizzly bear

Raven

Grizzly bear

Coho salmon

Sitka spruce

Red elderberry

You can use a **nested circle diagram** to show parts within a whole.

Grizzly bears are a part of a forest community.

To Compare and Contrast

Comparison of Methods of Plant Reproduction

You can use a **comparison matrix** to record and compare observations or results.

Plant	Seeds from cones	Spores	Seeds from flowers
pine tree	X		
fern			X
dandelion		X	

Comparing Plant and Animal Cells

You can use a **Venn diagram** to show similarities and differences. Similarities go in the middle section.

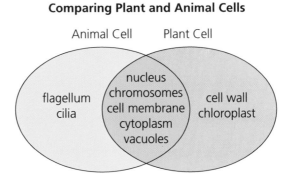

Animal Cell   Plant Cell

flagellum
cilia

nucleus
chromosomes
cell membrane
cytoplasm
vacuoles

cell wall
chloroplast

Plant and Animal Cells

You can use a **compare and contrast chart** to show both similarities and differences.

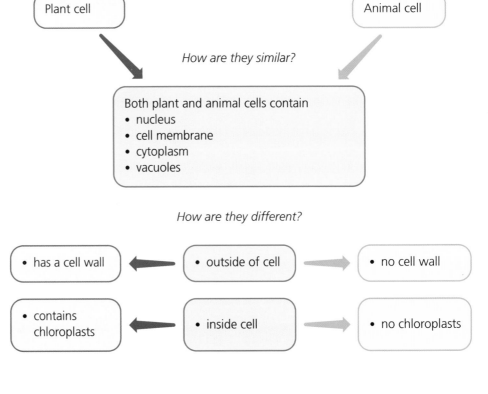

Plant cell

Animal cell

How are they similar?

Both plant and animal cells contain
- nucleus
- cell membrane
- cytoplasm
- vacuoles

How are they different?

- has a cell wall ← - outside of cell → - no cell wall

- contains chloroplasts ← - inside cell → - no chloroplasts

To Show Properties or Characteristics

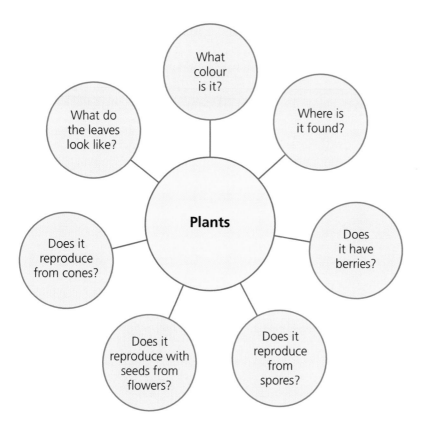

You can use a **bubble map** to show properties.

READING STRATEGIES

The skills and strategies you use to help you read can differ, depending on the type of material you are reading. Reading a science text is different from reading a novel. When you are reading a science text, you are reading for information. Here are some strategies to help you read for information.

Before Reading

Skim the section you are going to read. Look at the illustrations, headings, and subheadings.

- *Preview.* What is this section about? How is it organized?

- *Make connections.* What do I already know about the topic? How is it connected to other topics I have already learned?

- *Predict.* What information will I find in this section? Which parts will give me the most information?

- *Set a purpose.* What questions do I have about the topic?

During Reading

Pause and think as you read. Spend time on the photographs, illustrations, tables, and graphs, as well as on the words.

- *Check your understanding.* What are the main ideas in this section? How would I explain them in my own words? What questions do I still have? Do I need to reread? Do I need to read more slowly, or can I read more quickly?

- *Determine the meanings of key science terms.* Can I figure out the meaning of unfamiliar terms from context clues in words or illustrations? Do I understand the definitions of terms in bold type? Is there something about the structure of a new term that will help me remember its meaning? Are there terms I should look up in the glossary?

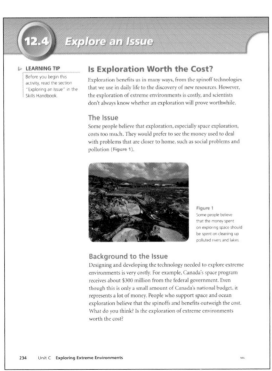

- *Make inferences.* What conclusions can I make from what I am reading? Can I make any conclusions by "reading between the lines"?

- *Visualize.* What mental pictures can I make to help me understand and remember what I am reading? Would it help to make a sketch?

- *Make connections.* How is this like things I already know?

- *Interpret visuals and graphics.* What additional information can I get from the photographs, illustrations, charts, or graphs?

After Reading

Many of the strategies you use during reading can be used after reading as well. For example, in this text, there are questions to answer after you read. These questions will help you check your understanding and make connections.

- *Locate needed information.* Where can I find the information I need to answer the questions? Under what heading might I find the information? What terms in bold type should I skim for? What details do I need to include in my answers?

- *Synthesize.* How can I organize this information? What graphic organizer could I use? What headings or categories could I use?

- *React.* What are my opinions about this information? How does it, or might it, affect my life or my community? Do other students agree with my reactions?

- *Evaluate information.* What do I know now that I did not know before? Have any of my ideas changed as a result of what I have read? What questions do I still have?

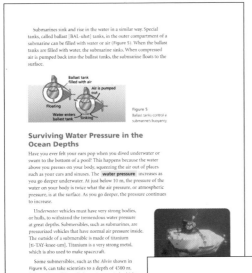

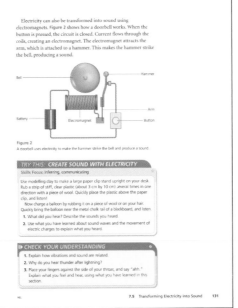

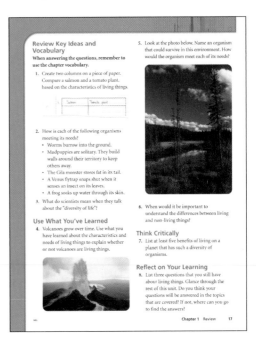

RESEARCHING

There is an incredible amount of scientific information that is available to you. Here are some tips on how to gather scientific information efficiently.

- Identify the Information You Need
- Find Sources of Information
- Evaluate the Sources of Information
- Record and Organize the Information
- Communicate the Information

Identify the Information You Need

Identify your research topic. Identify the purpose of your research.

Identify what you, or your group, already know about your topic. Also identify what you do not know. Develop a list of key questions that you need to answer. Identify categories based on your key questions. Use these categories to identify key search words.

Find Sources of Information

Identify all the places where you could look for information about your topic. These places might include videotapes of science programs on television, people in your community, print sources (such as books, magazines, and newspapers), and electronic sources (such as CD-ROMs and Internet sites). The sources of information might be in your school, home, or community.

Evaluate the Sources of Information

Preview your sources of information, and decide whether they are useful. Here are four things to consider.

- *Authority:* Who wrote or developed the information or sponsors the Web site? What are their qualifications?
- *Accuracy:* Are there any obvious errors or inconsistencies in the information? Does the information agree with other reliable sources?
- *Currency:* Is the information up to date? Has recent scientific information been included?
- *Suitability:* Does the information make sense to someone your age? Do you understand it? Is it organized in a way that you understand?

Record and Organize the Information

Identify categories or headings for note taking. Record information, in your own words, under each category or heading, perhaps in point form. If you quote a source, use quotation marks.

Record the sources to show where you got your information. Include the title, author, publisher, page number, and date. For Web sites, record the URL (Web site address).

If necessary, add to your list of questions as you find new information.

Communicate the Information

Choose a format for communication that suits your audience, your purpose, and the information.

COMMUNICATING IN SCIENCE

CREATING DATA TABLES

Data tables are an effective way to record both qualitative and quantitative observations. Making a data table should be one of your first steps when conducting an investigation. You may decide that a data table is enough to communicate your data, or you may decide to use your data to draw a graph. A graph will help you analyze your data. (See "Graphing Data," on page 272, for more information about graphs.)

Sometimes you may use a data table to record your observations in words, as shown below in the classification key set up as a table.

Observations from Investigation 7.2

Plant	Number of leaves on stem	Position of leaf on stem	Shape of leaf	Vein pattern of leaf	Size of leaf	Colour of leaf	Texture of leaf

Sometimes you may use a data table to record the values of the independent variable (the cause) and the dependent variables (the effects), as shown to the left. (Remember that there can be more than one dependent variable in an investigation.)

Average Monthly Temperatures in Cities A and B

Month	Temperature (°C) in City A	Temperature (°C) in City B
January	-7	-6
February	-6	-6
March	-1	-2
April	6	4
May	12	9
June	17	15

Follow these guidelines to make a data table:

- Use a ruler to make your table.
- Write a title that describes your data as precisely as you can.
- Include the units of measurements for each variable, when appropriate.
- List the values of the independent variable in the left-hand column of your table.
- List the values of the dependent variable(s) in the column(s) to the right of the independent variable.

GRAPHING DATA

When you conduct an investigation or do research, you often collect a lot of data. Sometimes the patterns or relationships in the data are difficult to see. For example, look at the data in **Table 1**.

Table 1 Average Rainfall in Campbell River

Month	Rainfall (mm)
January	142
February	125
March	128
April	73
May	59
June	50
July	40
August	43
September	62
October	154
November	210
December	197

One way to arrange your data so that it is easy to read and understand is to draw a graph. A graph shows numerical data in the form of a diagram. There are three kinds of graphs that are commonly used:

- bar graphs
- line graphs
- circle (pie) graphs

Each kind of graph has its own special uses. You need to identify which type of graph is best for the data you have collected.

Bar Graphs

A **bar graph** helps you make comparisons and see relationships when one of two variables is in numbers and the other is not. The following bar graph was created from the data in **Table 1**. It clearly shows the rainfall in different months of the year and makes comparison easy.

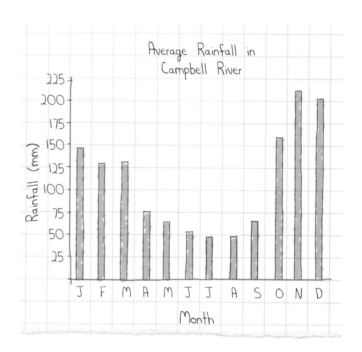

Line Graphs

A **line graph** is useful when you have two variables in numbers. It shows changes in measurement. It helps you decide whether there is a relationship between two sets of numbers: for example, "if this happens, then that happens." **Table 2** gives the number of earthworms found in specific volumes of water in soil. The line graph for these data helps you see that the number of earthworms increases as the volume of water in soil increases.

Table 2 Number of Earthworms per Volume of Water in Soil

Volume of water in soil (mL)	Number of earthworms
0	3
10	4
20	5
30	9
40	22

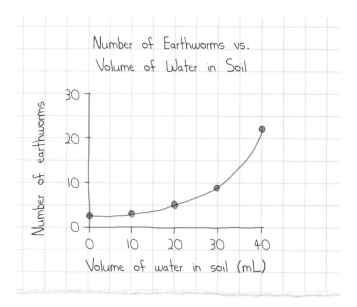

Circle Graphs

A **circle graph** (or pie graph) shows the whole of something divided into all its parts. A circle graph is round and shows how large a share of the circle belongs to different things. You can use circle graphs to see how the different things compare in size or quantity. It is a good way to graph data that are percentages or can be changed to percentages.

WRITING A LAB REPORT

When you design and conduct your own experiment, it is important to report your findings. Other people may want to repeat your experiment, or they may want to use or apply your findings to another situation. Your write-up, or report, should reflect the process of scientific inquiry that you used in your experiment.

Write the title of your experiment at the top of the page.

List the question(s) you were trying to answer. This section should be written in sentences.

Write your hypothesis. It should be a sentence in the form "If ... then ...."

Write the materials in a list. Your list should include equipment that will be reused and things that will be used up in the investigation. Give the amount or size, if this is important.

Describe the procedure using numbered steps. Each step should start on a new line and, if possible, it should start with a verb. Make sure that your steps are clear so that someone else could repeat your experiment and get the same results. Include any safety precautions.

Draw a large diagram with labels to show how you will set up the equipment. Use a ruler for straight lines.

Conductivity of Water

Question

Which type of water – pure water, water with dissolved sugar, or water with dissolved salt – conducts electricity the best?

Hypothesis

If water is very pure, like distilled water with no solutes, then it will conduct electricity better than water with sugar or salt dissolved in it.

Materials

3 clean glass jars	battery holder
distilled water	1 piece of wire, 25 cm long
sugar	2 pieces of wire, each 10 cm long
salt	wire strippers
3 short strips of masking tape	light-bulb holder
pen	small light bulb (such as a flashlight bulb)
2 D-cell batteries	

Procedure

1. Put 250 mL of distilled water in each clean jar. Do not add anything to the first jar. Add 30 mL of salt to the second jar, and mix. Add 30 mL of sugar to the third jar, and mix. Label the jars "pure water," "salt water," and "sugar water."
2. Put the batteries in the holder.
3. Strip the plastic coating off the last centimetre at the ends of all three wires, using the wire strippers.

CAUTION: Always pull the wire strippers away from your body.

4. Attach one end of the 25-cm wire to the knobby end of the battery by tucking it in the battery holder. The other end of the wire should hang free for now.
5. Attach one end of a 10-cm wire to the flat part of the battery. Attach the other end to the clip in the light-bulb holder.

6. Place the light bulb in the holder.
7. Attach one end of the other 10-cm wire to the clip in the light-bulb holder. Let the other end hang free for now.
8. Dip the loose wire ends into the distilled water. Observe whether the light bulb goes on. Record "yes" or "no."
9. Repeat step 8 for the other two types of water.

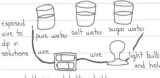

exposed wire to dip in solutions

pure water   salt water   sugar water

wire   wire   light bulb and holder

batteries and battery holder

Data and Observations

Type of water	Does the light bulb go on?
distilled water	no
water with salt	yes
water with sugar	no

Analysis
The salt water was the only type of water that turned on the light bulb. Something in the salt must help to conduct electricity. Since the distilled water did not turn on the light bulb, this must mean that it cannot conduct electricity. Something is missing from the distilled water. The sugar water did not conduct electricity either, so it must also be missing the ingredient that helps to conduct electricity.

Conclusion
Pure (distilled) water does not conduct electricity. The hypothesis is not supported by the data, so it is incorrect. Salt water conducts electricity.

Applications
Knowing that salt water conducts electricity might help scientists recover materials from seawater by running electricity through it. Also, I think the water in the human body has salt and other things dissolved in it. It would conduct electricity well, so people should be careful about electricity.

Present your observations in a form that is easily understood. The data should be recorded in one or more tables, with units included. Qualitative observations can be recorded in words or drawings. Observations in words can be in point form.

Interpret and analyze your results. If you have made graphs, include and explain them here. Answer any questions from the student text here. Your answers should include the questions.

A conclusion is a statement that explains the results of an experiment. Your conclusion should refer back to your hypothesis. Was your hypothesis correct, partly correct, or incorrect? Explain how you arrived at your conclusion. This section should be written in sentences.

Describe how the new information you gained from doing your experiment relates to real-life situations. How can this information be used?

GLOSSARY

A

amphibian a class of vertebrates that is born in water, but develops legs and can walk on land; examples include frogs, toads, and salamanders

Animalia part of the scientific system of classification; one of the five kingdoms; examples include insects, birds, fish, and mammals

atom the tiny building block that makes up everything around you; the air you breathe and the clothes you wear are made up of atoms

B

battery an energy source that uses a chemical reaction to create an electric current; is actually two or more electric cells connected together; term is commonly used to refer to one cell

behaviour the way an organism responds to its environment

biomass any type of plant or animal tissue, such as wood, straw, and crop waste; biomass can be burned to heat water and create steam to turn turbines and generate electricity

bird a class of vertebrates with feathers, wings, and a beak; examples include eagles, parrots, and cardinals

buoyancy the ability to float in water

C

camouflage the colouring of an animal that allows it to blend into its environment

Canadarm a robotic manipulator arm developed by the Canadian Space Agency that is controlled by astronauts inside the space shuttle

cell a microscopic structure that is the basic unit of all living things; organisms can be made of as little as one cell (some types of bacteria) or as many as several trillion cells (human beings)

cell membrane a thin covering around an entire cell that acts as a gatekeeper by controlling which materials move into and out of the cell

chlorophyll a green pigment found in chloroplasts that gives plants and some protists their green colour

chloroplast the cell structure containing chlorophyll; found in plant cells and some protists

circuit made of a source of electricity, a pathway, and an electrical device to operate; electric current flows around a complete circuit

classification system the organization of living or non-living things according to their similarities and differences

closed circuit a circuit that is complete; allows the current to flow along the pathway

coal a hard fossil fuel made of ancient plants such as trees and ferns

colouration an adaptation of an organism's colour to help it survive in its environment; mimicry and camouflage are examples of colouration

conductor a material that lets electricity flow through it easily; for example, metals are good conductors

conservation the careful and responsible use of energy resources; for example, turning out the light when you leave a room

consumption the amount of electricity used by a household; determined by meters placed on the transmission line that comes into your home

cover slip a small, thin piece of glass used to cover an object or specimen on a microscope slide

current electricity electricity produced by a flow of electrons through a conductor, such as a wire

D

direct current current that flows in one direction; a battery produces electric current

E

electric current the continuous flow of electrons from one place to another along a pathway

electrocute death caused by electric current

electromagnetism magnetic forces produced by electricity

electron a negatively charged particle that makes up an atom

environment your surroundings; other living organisms, non-living objects, and weather all make up your environment

exploration voyage into unknown territory to investigate new frontiers and to search for new discoveries

extreme harsh; beyond normal limits; to the greatest degree

extreme environment a place where the conditions are so harsh that human survival is difficult or impossible; for example, deserts, volcanoes, and space are extreme environments for humans

F

fish a class of vertebrates with gills and fins that lives in water; examples include salmon, whale sharks, and rays

fossil fuels the most common non-renewable energy sources used to produce electricity; made from the remains of dead organisms that lived millions of years ago; coal, natural gas, and oil are the three types of fossil fuels

Fungi part of the scientific system of classification; one of the five kingdoms; examples include yeast and mushrooms

G

geothermal energy energy from deep inside Earth that heats water and produces steam, which can then be used to turn turbines and produce electricity

H

hibernation a method of coping with winter where an animal's body temperature drops, and its heartbeat and breathing slow down; chipmunks and ground squirrels are animals that hibernate

hydro refers to hydroelectric energy, which is the electricity generated by the conversion of energy from moving water; accounts for approximately 85% of all electricity generated in British Columbia

hydroelectric dam a barrier that stops the flow of water on a river; an electricity generating station that converts the energy of moving water into electricity

I

indigenous knowledge understandings, values, and beliefs about the natural world that are unique to a particular group or culture who have lived for a very long time in a particular area; this specialized knowledge is passed from generation to generation in the form of stories told, experiences shared, or songs sung by Elders or other people

insulator a material that resists the flow of electricity (such as wood) and prevents heat from escaping (such as a winter jacket)

invertebrate an animal that does not have a backbone, or spinal column; examples of invertebrates include insects, worms, and crabs

K

kingdom the most basic grouping of all living things in the scientific system of classification (taxonomy); this text uses the five-kingdom system of classification—Plantae, Fungi, Animalia, Protista, and Monera

L

life-support system a system that helps astronauts to survive in space; for example, life-support systems allow astronauts to breathe easily in space where there is a lack of air

lift the movement of air around an airplane's wing creating an area of low pressure under the wing and an area of high pressure over the wing; the high pressure under the wing pushes the wing up and forces the airplane upward

light bulb electrical device that changes electricity into light and heat

M

magnetism the property of attracting or repelling iron

magnify to make objects appear larger, as with a lens

mammal a class of vertebrates that breathe oxygen from the air and are warm blooded; examples include bats, mice, and humans

micro-organism a very small living thing that can only be seen with a microscope

microscope a device used for viewing very small objects or specimens

migration the seasonal movement of animals to a less harsh environment; for example, the elk moves from the mountains to spend the winter in the lowlands

mimicry an adaptation where an organism looks like another organism

Monera part of the scientific system of classification; one of the five kingdoms; example includes bacteria

multicellular made of more than one cell; humans, for example, are multicellular

N

natural gas a fossil fuel that comes from plankton (tiny plants and animals) that lived in ancient seas and lakes; usually found with oil, often in deep wells

negative the kind of electric charge carried by electrons

non-renewable something that once used up cannot be replaced; coal is an example of a non-renewable resource

nuclear energy energy that uses uranium as a fuel to heat water and produce steam, which turns a turbine and produces electricity

nucleus the cell structure that acts as the control centre by directing all of the cell's activities, such as movement and growth

O

open circuit a circuit that is incomplete, current cannot flow along the pathway

organism a living thing, such as a plant or an animal

P

parallel circuit a type of circuit in which current can travel through more than one pathway

Plantae part of the scientific system of classification; one of the five kingdoms; examples include mosses, trees, and flowers

positive the kind of electric charge carried by protons

Protista part of the scientific system of classification; one of the five kingdoms; examples include algae and paramecia

R

radar an acronym for RAdio Detection And Ranging; a device that sends out radio waves and picks up any echoes that are bounced back off objects to tell the distance, speed, direction of motion, and shape of the object; boats and ships searching for land, ice, and other ships use radar

recycle to reuse something rather than discarding it; for example, water on the International Space Station is recycled

renewable something that is constantly being replaced and is always there to use; water is an example of a renewable resource

reptile a class of vertebrates that breathe through lungs and has a body temperature that depends on the external temperature; examples include crocodiles, alligators, and snakes

S

satellite an object in space that revolves around Earth or any other planet

scuba an acronym for Self-Contained Underwater Breathing Apparatus; allows divers to carry their air supply on their backs

series circuit a type of circuit in which current has only one pathway to travel through

slide a piece of glass that an object, or specimen, is placed on to be viewed under a microscope

solar energy   energy from the Sun; can be used to produce electricity by using mirrors to focus sunlight on water tanks and heat the water, producing steam which turns the turbines to generate electricity

sonar   an acronym for SOund NAvigation and Ranging; a device that ships use to chart the depth of oceans using the echoes of sound waves

species   organisms that are capable of breeding together and having fertile offspring

spinoff   an everyday use of a technology that was first developed for another purpose; for example, bar codes used in grocery stores were first developed by NASA for space exploration

static electricity   electricity where the electric charges are at rest, or not moving; caused by negative charges transferring from one object to another through rubbing; for example, electric charges built up in the clothes dryer or rubbing a balloon against your pet's fur

switch   a device used to control the flow of electric current

T

technology   tools that make it possible to survive in challenging environments

tidal energy   energy created by filling a reservoir with ocean water at high tide, and later releasing the water through hydroelectric turbines as the tide goes back out in order to produce electricity

thrust   an upward force

U

unicellular   made of only one cell; a characteristic of organisms in the Kingdom Monera and most organisms in the Kingdom Protista

V

vertebrate   animal with a backbone, or spinal column; birds, fish, and mammals are examples of vertebrates

voltage   the force or push that moves electrons in a circuit

W

water pressure   the application of force by water that increases with depth

wind power   energy created by wind pushing against the blades of a wind turbine and turning it, which then turns a magnet that generates electricity

INDEX

PHOTO CREDITS

Front Cover

Michael Patrick O'Neill/Photo Researchers, Inc.

Preface

Dr. Arif Babul, Department of Physics & Astronomy/
University of Victoria; Courtesy of Advanced Materials
& Process Engineering Laboratory/University of British
Columbia; Vancouver Aquarium Marine Science Centre;
Courtesy of Dennis Morgan/Bamfield Community Forest.

Table of Contents

p. vii Jerry Kobalenko/First Light; p. viii NASA;
p. ix Ralph White/Corbis Canada.

Unit A

Unit A Opener pp. 2–3 Jerry Kobalenko/First Light;
p. 4 Jeffrey L. Rotman/Corbis Canada; p. 6 left J.
Beveridge/Visuals Unlimited; right Jim Zuckerman/
Corbis Canada; p. 7 left Lynda Richardson/Corbis Canada;
right Ron Watts/Corbis Canada; p. 8 left J.A. Wilkinson/
VALAN Photos; right Chris Harris/First Light; p. 9 left
Brandon Cole/Visuals Unlimited; right W. Wayne
Lockwood/Corbis Canada; p. 12 top left Theo Allofs/
Corbis Canada; top right Jeffrey L. Rotman/Corbis Canada;
bottom left Joseph Sohm/Chromo Sohm Inc./Corbis
Canada; bottom right USDA/Visuals Unlimited; p. 13 top
left David Aubrey/Corbis Canada; top right Jeremy
Woodhouse/PhotoDisc/Getty Images; bottom Ken
Lucas/Visuals Unlimited; p. 14 left Matthew Wheeler;
top right Gavriel Jecan/Corbis Canada; bottom right
Stephen J. Kraseman/Photo Researchers, Inc.; p. 15 left
Michael & Patricia Fogden/Corbis Canada; top right
Courtesy of NOAA Fisheries Service; bottom right Martin
Harvey/Corbis Canada; p. 16 top left to right J. Beveridge/
Visuals Unlimited; Lynda Richardson/Corbis Canada;
Ron Watts/Corbis Canada; centre left to right
J.A. Wilkinson/VALAN Photos; Chris Harris/First Light;
Brandon Cole/Visuals Unlimited; W. Wayne Lockwood/
Corbis Canada; bottom clockwise top right to bottom left
Theo Allofs/Corbis Canada; Jeffrey L. Rotman/Corbis
Canada; Ken Lucas/Visuals Unlimited; David Aubrey/
Corbis Canada; p. 17 top Al Harvey; bottom Jim
Sugar/Corbis Canada; p. 20 © 2004 Mark Turner; p. 21
CDC/PHIL/Corbis Canada; p. 22 top left Brother Alfred
Brousseau/St. Mary's College of California; top right
Jo Kirchherr/StockFood/MaXx Images; bottom left Klaus
Arras/StockFood/MaXx Images; bottom right Virginia
Kline/University of Washington, Madison; p. 24 left

J. Harris/Ivy Images; centre Gerald & Buff Corsi/Visuals
Unlimited; right Kennan Ward/Corbis Canada; p. 26 left
Sheila Terry/Photo Researchers, Inc.; centre Bill
Beatty/Visuals Unlimited; right Courtesy of Dr. Nancy
Turner; p. 27 Peter Parks/Oxford Scientific Films; p. 28 top
Richard Kressel/Visuals Unlimited; bottom Cabisco/
Visuals Unlimited; p. 30 first row left to right T. Ulrich/
H. Armstrong Roberts/Comstock; Flip Nicklin/First Light;
First Light; second row left to right Ursus Photography;
Comstock; Geoff Kidd/Oxford Scientific Films; Rod
Planck/Photo Researchers; third row left to right Peter
Parks/Oxford Scientific Films; Grant Heilman/Comstock;
fourth row left to right T.E. Adams/Visuals Unlimited;
A.B. Dowsett/Science Photo Library; p. 31 Courtesy of
Burndy Library; p. 32 left to right Kelvin Aitken/First Light;
First Light; Comstock; Dag Sundberg/The Image Bank;
T. Ulrich/H. Armstrong Roberts/Comstock; p. 34 Hal
Beral/Visuals Unlimited; p. 37 top Ryan McVay/PhotoDisc/
Getty Images; bottom Corbis Canada; p. 38 Bob Gurr/
VALAN Photos; p. 39 left Dr. John D. Cunningham/
Visuals Unlimited; right Janice Palmer; p. 41 top Biophoto
Associates/Photo Researchers, Inc.; bottom Gunter Marx
Photography; p. 43 top Jon Hinsch/Science Photo Library;
bottom M. Abbey/Photo Researchers, Inc.; p. 44 M. Kaleb/
Visuals Unlimited; p. 47 top left Kevin Schafer/Corbis
Canada; top right Joe McDonald/Corbis Canada; bottom
left D. Newman/Visuals Unlimited; bottom right Michel
Bourque/VALAN Photos; p. 48 Gary Gaugler/Visuals
Unlimited; p. 49 Jose Luis Pelaez, Inc./Corbis Canada;
p. 50 M. Abbey/Photo Researchers, Inc.; p. 51 left
Dr. Dennis Kunkel/Visuals Unlimited; right Jim
Zuckerman/Corbis Canada; p. 57 top left to right Clouds
Hill Imaging Ltd./Corbis Canada; M. Abbey/Photo
Researchers, Inc.; M.I. Walker/Photo Researchers, Inc.;
centre left to right M. Abbey/Photo Researchers, Inc.;
Tom Adams/Visuals Unlimited; bottom left to right
Carolina Biological/Visuals Unlimited; T.E. Adams/Visuals
Unlimited; p. 59 Dr. Dennis Kunkel/Visuals Unlimited;
p. 64 first row left to right M. Abbey/Photo Researchers,
Inc.; Dr. Dennis Kunkel/Visuals Unlimited; second row
left to right M.I. Walker/Photo Researchers, Inc.;
T.E. Adams/Visuals Unlimited; third row Jose Luis Pelaez,
Inc./Corbis Canada; p. 65 CDC/PHIL/Corbis Canada;
p. 66 John & Barbara Gerlach/Visuals Unlimited;
p. 67 James Watt/Visuals Unlimited; p. 68 top Raymond
Gehman/Corbis Canada; bottom Minden Pictures/
First Light; inset M. Harvey/MaXx Images; p. 69 top
John Gerlach/Visuals Unlimited; bottom ABPL/

Lanz Von Horsten/MaXx Images; p. 70 top Wayne Lankinen/VALAN Photos; bottom Yogi, Inc./Corbis Canada; p. 71 top William J. Weber/Visuals Unlimited; bottom Mike Severns/Tom Stack & Associates; p. 75 Joe McDonald/Visuals Unlimited; p. 76 Fritz Polking/ Visuals Unlimited; p. 77 top left Thomas Kitchin & Victoria Hurst/First Light; top right Tom Brakefield/Corbis Canada; bottom Arthur Morris/Visuals Unlimited; p. 78 Breck P. Kent/Animals Animals-Earth Scenes; p. 79 top Al Harvey; bottom Viola's Photo Visuals, Inc./Animals Animals-Earth Scenes; p. 80 left Gordon Fisher/MaXx Images; right Reuters/Corbis Canada; p. 82 top David A. Northcott/ Corbis Canada; bottom John McAnulty/Corbis Canada; p. 83 Andrew A. Bryant; p. 84 first row left to right Raymand Gehman/Corbis Canada; John Gerlach/Visuals Unlimited; Fritz Polking/Visuals Unlimited; second row left to right Reuters/Corbis Canada; Arthur Morris/Visuals Unlimited; Breck P. Kent/Animals Animals-Earth Scenes; bottom John McAnulty/Corbis Canada.

Unit B

Unit B Opener pp. 88–89 NASA; p. 90 Gene Rhoden/ Visuals Unlimited; p. 96 Roger Ressmeyer/Corbis Canada; p. 97 top Michael Newman/PhotoEdit; bottom Gene Rhoden/Visuals Unlimited; p. 102 bottom Roger Ressmeyer/Corbis Canada; p. 104 Ron Watts/Corbis Canada; p. 105 Mike Zens/Corbis Canada; p. 107 Corbis Canada; p. 117 left Digistock/First Light; right K. Handke/ Zefa/Masterfile; p. 118 top Mike Zens/Corbis Canada; p. 119 Design Pics/First Light; p. 120 Dr. Jeremy Burgess/ Science Photo Library; p. 121 left Design Pics/First Light; p. 122 Jeff J. Daly/Visuals Unlimited; p. 123 Francoise Sauze/Science Photo Library; p. 124 bottom A. Bartel/Publiphoto/Science Photo Library; p. 125 Yuriko Nakao/Reuters/Corbis Canada; p. 129 top Gaye Gerard Photography; bottom Al Harvey; p. 130 Roy Morsch/ Corbis Canada; p. 134 top Design Pics/First Light; centre left Roy Morsch/Corbis Canada; bottom A. Bartel/ Publiphoto/Science Photo Library; p. 136 Andrew Wallace/Reuters/Landov; p. 138 bottom Al Harvey; p. 140 Al Harvey; p. 142 Gary Braasch/Corbis Canada; p. 144 Mark Newman/Photo Researchers, Inc.; p. 145 N. Piluke/Ivy Images; p. 146 Reuters/Corbis Canada; p. 147 Courtesy of Meager Creek Development Corporation, www.geopower.ca; p. 148 Gunter Marx Photography/Corbis Canada; p. 149 Courtesy of Western Valley Development Agency; p. 150 Dawn Campbell/ Annapolis County; p. 151 Courtesy of City of Williams Lake Economic Development Office; p. 152 Al Harvey; p. 153 Larry Lee Photography/Corbis Canada; p. 154 Chris Falkenstein/PhotoDisc/Getty Images; p. 155 J.A. Wilkinson/VALAN Photos; p. 156 top Al Harvey;

bottom Courtesy of Joe Paradiso/MIT Media Lab; p. 158 second row Al Harvey; third row N. Piluke/Ivy Images; third row left to right Reuters/Corbis Canada; Courtesy of Meager Creek Development Corporation, www.geopower.ca; Courtesy of Western Valley Development Agency; Courtesy of City of Williams Lake Economic Development Office; bottom Chris Falkenstein/PhotoDisc/Getty Images; p. 159 Larry Lee Photography/Corbis Canada.

Unit C

Unit C Opener pp. 162–163 Ralph White/Corbis Canada; p. 164 Ann Hawthorne/Corbis Canada; p. 165 Les & Leslee Deacon-Rogers; p. 166 New Zealand Herald/Sygma/Corbis Canada; p. 168 Lowell Georgia/Corbis Canada; p. 169 top Gunter Marx Photography; bottom Steven Norvich/Visuals Unlimited; p. 170 Jim Sugar/Corbis Canada; p. 171 Allan E. Morton/Visuals Unlimited; p. 172 Ruth Gotthardt/Yukon Government; p. 173 top Stephen Frink/Corbis Canada; bottom Jeffrey L. Rotman/Corbis Canada; p. 174 Courtesy of Hibernia; p. 175 Courtesy of Save the Elephants, www.savetheelephants.org; p. 176 top B&C Alexander/ First Light; centre Alvaro Canovas/Sygma/Corbis Canada; bottom © 2004 Myron Wright/AlaskaStock.com; p. 177 top B&C Alexander/First Light; bottom B&C Alexander Photography; p. 178 Courtesy of Michael Schmidt; p. 181 Courtesy of Dr. Garry Clarke/University of British Columbia; p. 181 top New Zealand Herald/Sygma/Corbis Canada; second row left to right Lowell Georgia/Corbis Canada; Gunter Marx Photography; Steven Norvich/ Visuals Unlimited; Jim Sugar/Corbis Canada; Allan E. Morton/Visuals Unlimited; third row left to right Courtesy of Hibernia; Courtesy of Save the Elephants, www.savetheelephants.org; bottom left B&C Alexander/ First Light; bottom right Alvaro Canovas/Sygma/Corbis Canada; p. 184 Roger Ressmeyer/Corbis Canada; p. 188 left Imagestate/First Light; right A.G.E. Foto Stock/First Light; p. 193 top Stuart Westmorland/Corbis Canada; bottom Rick Gomez/Corbis Canada; p. 195 Woods Hole Oceanographic Institution; p. 196 Courtesy of Nuytco Research Ltd.; p. 197 top Neil McDaniel Photography; bottom Canadian Scientific Submersible Facility www.ropos.com; p. 198 top Ralph White/Corbis Canada; bottom Canadian Scientific Submersible Facility www.ropos.com; p. 200 third row left to right Stuart Westmorland/Corbis Canada; Woods Hole Oceanographic Institution; bottom left Courtesy of Nuytco Research Ltd.; bottom right Neil McDaniel Photography; p. 201 Canadian Scientific Submersible Facility www.ropos.com; p. 202 EPA/NASA/Landov; p. 203 Bettmann/Corbis Canada; p. 205 Bettmann/Corbis Canada; p. 206 left Corbis Canada; centre Michael & Patricia Fogden/Corbis Canada;

right Dirk Sigmund/Picture Press/First Light; p. 208 Esther C. Goddard/AIP Neils Bohr Library; p. 209 NASA; p. 210 Corbis Canada; p. 211 Courtesy of The Boeing Company; p. 212 top Courtesy of Canada Centre for Remote Sensing; bottom NASA; p. 213 NASA; p. 214 left NASA; right Bill Ingalls/NASA; p. 215 left Canadian Scientific Submersible Facility www.ropos.com; NASA; p. 219 NASA; pp. 220–222 NASA; p. 223 bottom NASA; pp. 224–225 NASA; p. 226 second row NASA; third row left to right NASA; Courtesy of Canada Centre for Remote Sensing; bottom NASA; p. 227 Mike Blake/Reuters/Corbis Canada; p. 229 top William Whitehurst/Corbis Canada; bottom left Ruth Jenkinson/MIDIRS/Photo Researchers, Inc.; bottom right Pitchel Frederic/Sygma/Corbis Canada; p. 231 NASA; p. 232 top NASA; bottom Detler Van Ravenswaay/Science Photo Library; p. 233 NASA/Science Photo Library; p. 234 Michael St. Maur Shiel/Corbis Canada; p. 236 bottom Detler Van Ravenswaay/Science Photo Library; p. 237 NASA; p. 238 Chris McLaughlin/Corbis Canada; p. 239 Courtesy of Jack Jones/Jet Propulsion Laboratory.

Skills Handbook

p. 248 bottom Courtesy of Boreal; p. 249 Rick Fischer/ Masterfile; p. 256 Lyle Ottenbreit; p. 262 top left Boreal; bottom left Richard L. Carlton/Visuals Unlimited; p. 270 top left Todd Ryoji; bottom left Omni Photo Communications/Index Stock.

Back Cover

Top Jerry Kobalenko/First Light; middle NASA; bottom Ralph White/Corbis Canada

Additional Photography

Ray Boudreau
Dave Starrett